ABOUT ISLAND PRESS

Island Press is the only nonprofit organization in the United States whose principal purpose is the publication of books on environmental issues and natural resource management. We provide solutions-oriented information to professionals, public officials, business and community leaders, and concerned citizens who are shaping responses to environmental problems.

In 2004, Island Press celebrates its twentieth anniversary as the leading provider of timely and practical books that take a multidisciplinary approach to critical environmental concerns. Our growing list of titles reflects our commitment to bringing the best of an expanding body of literature to the environmental community throughout North America and the world.

Support for Island Press is provided by the Agua Fund, Brainerd Foundation, Geraldine R. Dodge Foundation, Doris Duke Charitable Foundation, Educational Foundation of America, The Ford Foundation, The George Gund Foundation, The William and Flora Hewlett Foundation, Henry Luce Foundation, The John D. and Catherine T. MacArthur Foundation, The Andrew W. Mellon Foundation, The Curtis and Edith Munson Foundation, National Environmental Trust, National Fish and Wildlife Foundation, The New-Land Foundation, Oak Foundation, The Overbrook Foundation, The David and Lucile Packard Foundation, The Pew Charitable Trusts, The Rockefeller Foundation, The Winslow Foundation, and other generous donors.

The opinions expressed in this book are those of the author(s) and do not necessarily reflect the views of these foundations.

ABOUT THE SOCIETY FOR ECOLOGICAL
RESTORATION INTERNATIONAL

The Society for Ecological Restoration International is an international nonprofit organization composed of members who are actively engaged in ecologically sensitive repair and management of ecosystems through an unusually broad array of experience, knowledge sets, and cultural perspectives.

The mission of SER International is to promote ecological restoration as a means of sustaining the diversity of life on Earth and reestablishing an ecologically healthy relationship between nature and culture.

SER International, 1955 W. Grant Road, Suite 150, Tucson, AZ 85745. Tel. (520) 622-5485, Fax (520) 622-5491, E-mail info@ser.org, www.ser.org.

GREAT BASIN RIPARIAN AREAS

SOCIETY FOR ECOLOGICAL RESTORATION INTERNATIONAL

The Science and Practice of Ecological Restoration
James Aronson, EDITOR
Donald A. Falk, ASSOCIATE EDITOR

Wildlife Restoration: Techniques for Habitat Analysis and Animal Monitoring, by Michael L. Morrison

Ecological Restoration of Southwestern Ponderosa Pine Forests, edited by Peter Friederici and Ecological Restoration Institute at Northern Arizona University

Ex Situ Plant Conservation: Supporting Species Survival in the Wild, edited by Edward O. Guerrant Jr., Kayri Havens, and Mike Maunder

Great Basin Riparian Areas: Ecology, Management, and Restoration, edited by Jeanne C. Chambers and Jerry R. Miller

Great Basin Riparian Areas

Ecology, Management, and Restoration

Edited by
Jeanne C. Chambers and Jerry R. Miller

Foreword by James A. MacMahon

SOCIETY FOR ECOLOGICAL RESTORATION INTERNATIONAL

ISLAND PRESS

Washington · Covelo · London

Library of Congress Cataloging-in-Publication data.

Great Basin riparian areas : ecology, management, and restoration / edited by Jeanne C. Chambers and Jerry R. Miller : foreword by James A. MacMahon.
    p. cm. — (Science and practice of ecological restoration)
Includes bibliographical references (p.).
    ISBN 1-55963-986-5 (cloth : alk. paper) — ISBN 1-55963-987-3 (pbk. : alk. paper)
    1. Riparian ecology—Great Basin. 2. Riparian restoration—Great Basin. I. Chambers, Jeanne C. II. Miller, Jerry R. (Jerry Russell), 1960– III. Series.
    QH104.5.G68G74 2004
577.68—dc22

2003024787

British Cataloguing-in-Publication data available.

Printed on recycled, acid-free paper ♻

Design by Teresa Bonner

Manufactured in the United States of America

10 9 8 7 6 5 4 3 2 1

CONTENTS

The Great Basin is big and diverse. The physiographic boundaries of the basin encompass over 650,000 square kilometers of land that occupies most of the state of Nevada, the western third of Utah, as well as small portions of California, Idaho, Oregon, and Wyoming. The Rocky Mountains to the east and the Sierra Nevada to the west form the most obvious boundaries to the basin. The entire basin is dotted with more than 200 mountain ranges containing over 500 mountains that often attain elevations in excess of 3300 meters, with a few greater than 4000 meters. The vegetation of the Great Basin is as diverse as its topography, varying from barren salt flats fringed with a few highly adapted, salt-tolerant plants, through an "ocean of sagebrush," to a variety of coniferous forests, and upward past treeline to areas of starkly beautiful alpine vegetation. The landscape is dotted with over seventy basins that receive the input of rivers and streams from the surrounding mountains. Most basins have no outlets, so ephemeral lakes, dry most of the time, form. Interspersed in this landscape are ribbons of riparian vegetation that, while occupying only 1 percent of the land surface, are inordinately important for sustaining the biodiversity of plants and animals, as well as for humans because water is such a scarce commodity across the Great Basin.

These riparian areas are so important that they have been extensively and intensively used for decades by humans for a variety of purposes that range from providing well-vegetated sites for grazing to places of beauty and solace that renew the spirits of visitors.

While the Great Basin has a history of occupation by Native Americans, their populations were not numerous. The landscape escaped many of the effects of early European settlement because the land was considered barren and, thus, not useful for agriculture. This began to change in the mid-1800s when the Mormon pioneers arrived. Today various municipalities in the Great Basin are among the fastest growing in the United States, the little available farmland is being subdivided for housing, and the constant call for more water continues. The past and present uses of riparian areas has left marks that signal a degree of degradation that suggests that the land cannot continue to support sustainably the inevitable increase of human activities into the future.

The future scenario is even more complicated because of the prospects global warming carries with it, including the possibility not only of increasing temperatures, but, more important, the seasonality and total annual precipitation. Even the dominant forms of precipitation (water versus snow) may change. Some of these changes may be positive, however, many are predicted to have negative consequences for the population as a whole, and they may have more pervasive effects on the human lifestyles of the Great Basin than in many other parts of the country.

How do we stem the tide of degradation and renew or recreate these systems so they can continue to support a reasonable human population, as well as their constituent plants and animals, as they have in the past? This is the topic of *Great Basin Riparian Areas: Ecology, Management, and Restoration*, an important and timely book that emanates from a research program funded by the U.S. Forest Service that commenced in 1993. The project took a fresh approach to addressing problems of riparian ecosystem restoration and management. Instead of jumping right into a restoration project based on past practices, as is often the case in the United States, this group of scientists decided to attempt to understand the natural functioning of riparian systems, including all the usual biological processes, but with unusual attention to the physical processes associated with the system's geomorphology and hydrology. The result of this research protocol is the knowledge needed to develop effective restoration and management plans, based on sound principles derived from attention to natural systems. This approach, in my opinion, is the way to do restoration science. Rather than creating schemes that rely only on principles derived from the applied sciences, such as agriculture or engineering, we need to learn how a system came to be, how it functions, and how it persists. As the natural systems reveal their intricate inner workings to us, then and only then can we develop the mechanistic plans to restore or manage them in the most efficient ways, often saving time and money in the process.

The eighteen authors of this book, under the guidance of Jeanne C. Chambers and Jerry R. Miller, have produced a volume that represents a substantial contribution, is packed with innovative thinking, and addresses not just the surprisingly limited literature on the Great Basin but, more importantly, includes consideration of the burgeoning recent literature on restoration elsewhere. Surely, Great Basin riparian zones will be more persistent and productive because of this careful research and synthesis. Of even greater potential significance, the entire field of restoration will be better because its practitioners are sure to adopt much of the general approach and many of the insights presented here as they tackle restoration problems in a wide variety of vegetation types, not just those in the arid American West.

This team took the time to listen to the lessons natural systems offer. Their example is one restorationists and land managers should all follow. I know I will and I thank them for their hard work, careful thought, and inspiration.

James A. MacMahon
Trustee Professor of Ecology
Department of Biology and the Ecology Center
Utah State University

In 1993, the USDA Forest Service initiated a series of Ecosystem Management Projects across the United States. When the supervisor of the Humboldt-Toiyabe National Forest (1.3 million acres that encompass most of the national forest lands within the Great Basin) was asked what was the Forest's priority for research within an ecosystem management context, he immediately replied, "riparian areas." For much of their history, the major land management agencies in the region, the Forest Service and the Bureau of Land Management, have largely treated riparian areas as sacrificial. Many were adversely affected by stream diversions or water-spreading, degraded by road systems, and severely overgrazed by livestock. The passage of a series of legislative acts in the mid-1970s, however, including the Forest and Rangeland Renewable Resources Planning Act, the National Forest Management Act, and the Endangered Species Act, drew needed attention to these valuable but often neglected ecosystems. The role of riparian ecosystems in supplying important commodities, such as livestock forage, had always been recognized, but now their role in providing other critically important ecosystem services, such as an adequate source of high-quality water and habitat for a diverse array of both terrestrial and aquatic organisms, was receiving increasing attention.

With the support of both the Humboldt-Toiyabe National Forest and the research branch of the Forest Service, in this case the Rocky Mountain Research Station, the Great Basin Ecosystem Management Project for Restoring and Maintaining Sustainable Riparian Ecosystems (EM Project) was funded. The Project was structured as a collaborative effort between management and research, and initial meetings focused on the overall direction and objectives of the project. Some objectives and approaches received complete agreement. For example, all agreed that permanent stream cross sections, wells, and vegetation transects should be established in key riparian ecosystems to determine existing ecological conditions and to monitor change over time. Other objectives were more difficult to decide upon. Some individuals wanted to begin restoration activities immediately, and it was suggested that contractors be hired to lay back the banks of incised streams and realign stream channels to their "predisturbance" pattern and form. Others argued, however, that without an understanding of the geomorphic and hydrologic processes influencing the streams and their relationships to riparian vegetation, these types of restoration efforts could fail. Several managers thought that a major focus of the project should be on the effects of overgrazing by livestock, a widespread problem in the Great Basin. A review of the literature on riparian areas in the semiarid regions of the western United States indicated that a number of studies had been conducted on the effects of livestock on stream channels and riparian ecosystems. However, study results depended on stream attributes, hydrologic regimes, and vegetation characteristics, and, again, conflicting findings were attributed to a lack of understanding of underlying processes (Trimble and Mendel 1995; Belsky et al. 1999). Ultimately, it was decided that to develop viable management and restoration approaches for riparian areas in the Great Basin, it would be necessary to first acquire an understanding of the interrelated responses of geomorphic, hydrologic, and biotic processes to both natural and anthropogenic disturbances. Because semiarid ecosystems like the Great Basin are highly sensitive to climate change, effective restoration and management also would require an understanding of how these processes are affected by past and present climates. Finally, in the context of ecosystem management, knowledge of these processes would have to be obtained over a continuum of temporal and spatial scales. Although not the major focus of the EM Project, it has been possible to examine the effects of overgrazing by livestock on key riparian ecosystems. And the knowledge obtained on the geomorphic, hydrologic, and biotic processes that characterize these ecosystems has allowed us to develop process-based restoration approaches that now are being put into practice.

This book presents the approach used by the EM Project to study and understand Great Basin riparian areas. It attempts to summarize the current state of knowledge about these areas and to provide insights into the use of this information for the restoration and management of riparian ecosystems. There are many ways to structure ecosystem management projects, and questions still remain regarding the management and restoration of Great Basin riparian areas. Nonetheless, we hope that both the approaches described and the information provided will serve to stimulate and guide future research on the topic.

Many individuals have made significant contributions to the EM Project. We thank M. Dean Knighton (Rocky Mountain Research Station, now retired) for his help in developing the EM Project and Larry J. Schmidt (director, Stream Systems Technology Center) for his technical guidance as well as financial support. Humboldt-Toiyabe National Forest staff were instrumental in conducting all aspects this work: Jim Nelson and Bob Vaught (past and present forest supervisors), Dale Flannigan and Laurence Crabtree (past and present district rangers), Jerry Grevstad, Ron Burraycheck, Dave Weixelman, Desi Zamudio, Karen Zamudio, Randy Sharp (Ecology Team), Jim Bergman (forest hydrologist), Dave Haney and Stuart Volkland (Fire Management), Terry Nevius, Bryan Watts, Will Wilson, Wayne and Peggy Frye, Joe and Christy Shaw, Liz Bergey, and Michael Croxen (Austin Ranger District personnel), and many others. The University of Nevada, Reno, provided administrative support and helped eight graduate students complete their degrees: David Martin, Amy Linnerooth, Regine Castelli, Catherine Davis, Danielle Henderson, Pam Wehking, Michael Wright, and Dan Lahde. Finally, we thank the EM Project's research collaborators (most are included in the section About the Editors and Authors) who have shared the project's meager budget, contributed their own time and finances, and helped to obtain additional resources to continually address "the next set of critical questions."

James Aronson and Don Falk, the editors of the SERI book series, supported the project from its beginnings, and Barbara Dean and Laura Carrithers of Island Press guided us through the process of turning a collection of papers into a book volume.

Literature Cited

Belsky, A. J., A. Matzke, and S. Uselman. 1999. Survey of livestock influences on stream and riparian ecosystems in the western United States. *Journal of Soil and Water Conservation* 51:419–431.

Trimble, S. W., and A. C. Mendel. 1995. The cow as a geomorphic agent: A critical review. *Geomorphology* 13:233–253.

Chapter 1

Restoring and Maintaining Sustainable Riparian Ecosystems: The Great Basin Ecosystem Management Project

JEANNE C. CHAMBERS AND JERRY R. MILLER

In the Great Basin, as in other semiarid regions, riparian areas exhibit widespread degradation. It has been estimated that more than 50 percent of the riparian areas (streams and their associated riparian ecosystems) in the Great Basin are currently in poor ecological condition (Jenson and Platts 1990). The ongoing deterioration of these areas is of significant concern to land managers and other stakeholders who value these watersheds for a variety of purposes. Riparian areas are important components of all landscapes, but in the semiarid Great Basin they constitute an especially vital resource. Although they comprise less than 1 percent of the Great Basin, they supply many critical ecosystem services. Riparian areas supply water for both culinary and agricultural uses, forage and browse for native herbivores and livestock, and recreational opportunities. In addition, they serve as the foundation for much of the region's biodiversity. Riparian areas in the Great Basin provide habitat for a wide array of organisms such as butterflies (Fleishman et al. 1999) and Neotropical migrant birds (Martin and Finch 1996), and support a relatively high number of endemic species, including the Lahontan cutthroat trout (*Oncorhynchus clarki henshawi*), which is listed as threatened under the U.S. Endangered Species Act (Dunham et al. 1997).

Degradation of riparian areas in the Great Basin is the result of complex and interrelated responses of geomorphic, hydrologic, and biotic processes to climate change and natural and anthropogenic disturbances. These disturbances can alter the hydrologic or sedimentologic regime of a fluvial (river or stream) system and produce changes in the physical foundations of riparian ecosystems, such as stream channel characteristics and surface-groundwater interactions. Ultimately, they alter the structure and functioning of riparian ecosystems. In this volume, anthropogenic disturbances

1

refer to all human activities that affect physical and biological processes within a watershed, while natural disturbances include phenomena such as floods, landslides, and wildfires. Although climate change could be considered to be a type of natural disturbance, it operates over longer temporal scales and larger spatial scales than most other forms of natural disturbance. Also, current shifts in climate arguably are related to human activities. Thus, climate change is treated as a special form of disturbance herein.

Much of the research on stream and riparian ecosystem degradation in arid and semiarid regions of the western United States has focused on the effects of anthropogenic disturbances. Consequently, the degradation of these riparian areas has been attributed largely to human activities, and management and restoration strategies have focused primarily on anthropogenic disturbances. In the Great Basin, riparian areas and their associated uplands have been subjected to various anthropogenic disturbances since European settlement of the region around 1860. The most extensive disturbances have been overgrazing by livestock (Kauffman and Krueger 1984; Fleischner 1994; Ohmart 1996; Trimble and Mendel 1995; Belsky et al. 1999) and road construction in the valley and canyon bottoms. Local alterations of hydrologic regimes via dams and water diversions, mining operations, and recreational activities also have had negative influences on riparian ecosystems (Sidle and Amacher 1990; Sidle and Hornbeck 1991). The influences of these disturbances on riparian areas have been well documented and management strategies for mitigating their effects are discussed in numerous locations (e.g., Kusler and Kentula 1990; National Research Council 1992, 2002; Briggs 1996; Kauffman et al. 1997; Williams et al. 1997).

The effects of past and present climate change on stream and riparian ecosystems have received considerably less attention than the influences of anthropogenic disturbance. This is surprising given that arid and semiarid regions like the Great Basin are more sensitive to the effects of both past and present climate change than humid regions. The sensitivity of these regions to climate change has important implications for the types and characteristics of disturbances that riparian areas experience, and the effects of these disturbances on riparian ecosystems. In comparison to humid regions, arid and semiarid regions generally exhibit amplified runoff responses to precipitation change (Dahm and Molles 1991), have higher streamflow variability (Poff 1991; Osterkamp and Friedman 2000), and have more severe flash floods (Graf 1988; Osterkamp and Friedman 2000).

In the Great Basin, paleoecological and geomorphic records indicate that there have been significant fluctuations in climate during the Holocene (approximately the past ten thousand years) (Tausch and Nowak 2000), and that these fluctuations have had major effects on disturbance regimes (Miller et al. 2001). Changes in hillslope processes, stream channel pattern and form, surface and groundwater interactions, and riparian vegetation composition and structure over time scales of hundreds of years have all been attributed to Holocene shifts in climate (Chambers et al. 1998; Miller et al. 2001). Perhaps more important from a management and restoration perspective is that the effects of these changes on hillslope processes and landforms have persisted for hundreds to thousands of years. For example, a shift from moister to drier conditions during the mid- to late-Holocene led to accelerated hillslope erosion, sediment deposition on alluvial fans and in valley bottoms, and a depletion of hillslope sediment supplies in upland watersheds of the central Great Basin (Miller et al. 2001). These climate-induced changes still influence geomorphic processes and, thus, channel pattern and form.

The failure of past restoration activities in semiarid riparian areas to meet desired goals has been attributed to a general lack of understanding of existing physical and biotic processes and the causes of disturbance (Elmore and Kauffman 1994; Kauffman et al. 1997; Goodwin et al. 1997), and to the use of small-scale, site-specific approaches that fail to consider watershed scale processes (Roper et al. 1997; Williams et al. 1997). Clearly, developing appropriate management and restoration approaches for riparian areas in the Great Basin requires an understanding of the responses of geomorphic, hydrologic, and biotic processes not only to natural and anthropogenic disturbances, but also to climate change. It also requires knowledge of these processes over sufficiently large temporal and spatial scales.

The Great Basin Ecosystem Management Project

In 1992, a USDA Forest Service Research ecosystem management project, "Restoring and Maintaining Sustainable Riparian Ecosystems," was initiated to address the problem of stream and riparian ecosystem degradation within the central Great Basin. The Great Basin Ecosystem Management (EM) Project uses an integrated, interdisciplinary approach to increase our understanding of the effects of climate change and anthropogenic disturbance on riparian areas, and to elucidate the connections among watershed and channel processes, hydrologic regimes, and riparian ecosystem dynamics. The EM Project is unique in that it addresses temporal scales ranging from the mid-Holocene to the present and spatial scales ranging from entire watersheds to localized stream reaches. The project's process-based and multiscaled approach is used to develop guidelines and methods for maintaining and restoring sustainable riparian ecosystems.

Definitions and Concepts

Ecosystem management uses ecological, economic, social, and managerial principles to maintain, restore, or create ecosystems that are capable of sustaining desired uses, products, values, and services over long time periods (modified from Overbay 1992). Thus, restoration is an integral component of contemporary ecosystem management. The Society for Ecological Restoration International's (SERI) definition of restoration is consistent with the concepts inherent in ecosystem management. Ecological restoration is defined by SERI as the process of assisting the recovery of an ecosystem that has been degraded, damaged, or destroyed (Society for Ecological Restoration International 2002). Ecosystem management has focused on watershed and regional scales, regardless of ecological condition, and emphasized the need to include both larger spatial and longer temporal scales. In contrast, restoration ecology has focused on degraded, damaged, or destroyed ecosystems but increasingly recognizes the need to consider larger scales in developing viable restoration approaches (Naveh 1994; Hobbs and Harris 2001). At the core of both ecosystem management and restoration ecology is the concept of sustainability. Sustainable ecosystems, over the normal cycle of disturbance events, retain characteristic processes including hydrologic flux and storage, geomorphic processes, biogeochemical cycling and storage, and biological activity and production (modified from Chapin et al. 1996 and Christensen et al. 1996). Because riparian areas serve as the interface between upland and stream ecosystems, sustainable stream and riparian ecosystems exhibit physical, chemical, and biological linkages among their geomorphic, hydrologic, and biotic components (Gregory et al. 1991). Thus, for the purposes of this volume, managing and restoring riparian areas is defined as maintaining or reestablishing sustainable fluvial systems and riparian ecosystems that exhibit both characteristic processes and related biological, chemical, and physical linkages among system components (modified from National Research Council 1992). Inherent in this definition is the notion that sustainable ecosystems supply important ecosystem services.

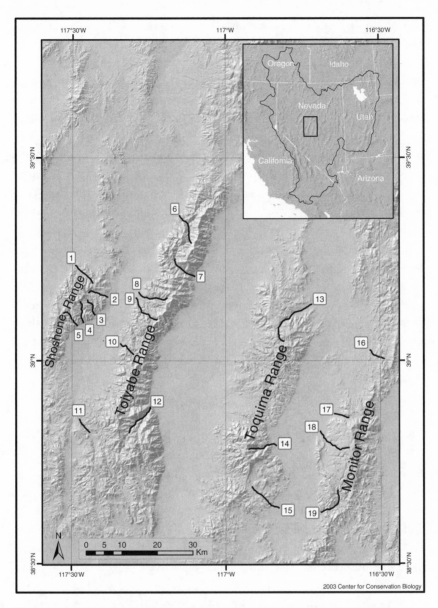

FIGURE 1.1. The distribution of the Great Basin in the western United States, and the locations of selected study watersheds in central Nevada: (1) Schoonnorer, (2) Underdown, (3) Riley, (4) Becker, (5) Barrett, (6) Big Creek, (7) Kingston, (8) Washington, (9) San Juan, (10) Marysville, (11) Upper Reese, (12) South Twin, (13) Stoneberger, (14) Pine, (15) Meadow Canyon, (16) Willow, (17) Morgan, (18) Mosquito, and (19) Barley.

To establish viable management and restoration goals, it is necessary to understand the current ecosystem or restoration potential. A frequent goal of both management and restoration has been to re-create and manage for the predisturbance condition (National Research Council 1992). In many cases, the predisturbance condition has been assumed to be the state of the stream or riparian ecosystem prior to European settlement. Several problems exist with this approach. First, we seldom understand the structure or processes of stream and riparian ecosystems prior to settlement. More important, using presettlement conditions as the goal of management or restoration assumes stable or equilibrium conditions over hundreds of years and ignores changes in stream processes and ecosystem dynamics due to changes in climate, hydrology, or land uses (Wade et al. 1998). A more realistic approach is to base management and restoration goals on the current potential of the streams and riparian ecosystems to support a given set of conditions. This approach requires an understanding of the types and magnitudes of biotic and abiotic thresholds that may have been crossed as a result of climate change or anthropogenic disturbance, and of the alternative states that currently exist for these systems (e.g., Hobbs and Norton 1996; Whisenant 1999; Hobbs and Harris 2001).

Objectives

The primary emphasis of the Great Basin EM Project has been on the geomorphic processes, hydrologic regimes, and vegetation dynamics of the systems. A parallel effort by the Nevada Biodiversity Initiative has examined the relationships among faunal distributions and the physical environment, and this volume includes an overview of various approaches for studying and managing faunal distributions in the Great Basin. The results of the EM Project that are presented in this volume were generated to accomplish the following series of interrelated objectives.

1. Reconstruct the vegetation and geomorphic history of central Great Basin watersheds to increase our understanding of the effects of past and present climate change on ecosystem processes.
2. Elucidate the underlying geomorphic and hydrologic processes that characterize the watersheds and riparian areas, and evaluate the effects of both past and present climate change and anthropogenic disturbance on these processes.

3. Determine the sensitivity of the study watersheds to both natural and anthropogenic disturbance, and develop a model of watershed sensitivity for managing riparian ecosystems.
4. Evaluate the effects of watershed geology and natural and anthropogenic disturbance on flow regimes and water quality, and relate this information to watershed sensitivity.
5. Determine the relationships among riparian vegetation dynamics and hydrogeomorphic processes, and evaluate the effects of past and present climate change and anthropogenic disturbance on those relationships.
6. Examine relationships among faunal distributions and the physical environment, and evaluate approaches for studying and managing native fauna, including documentation of historical changes.
7. Use our understanding of past and present ecosystem processes to develop guidelines and techniques for restoring and maintaining sustainable riparian areas.

Project Area

The EM Project is located in the central Great Basin (fig. 1.1). The Great Basin is the largest section of the Basin and Range Physiographic Province and contains almost all of the state of Nevada, the western half of Utah, and adjoining portions of Oregon, California, and Idaho. Although its name implies that the region consists of a single large depression, it is actually composed of a series of north-northwest-trending mountain ranges separated by alleviated, intermountain basins that were formed by high angle extensional faulting. In the central Great Basin, the ranges comprise more than 40 percent of the total landscape and may reach elevations in excess of 3,500 meters, more than 1,500 to 2,000 meters above the surrounding basin floors (Dohrenwend 1987). With the exception of a few areas along the north, northwestern, and southeastern margins of the Great Basin, drainage is completely internal. More than two hundred separate hydrologic systems can be delineated, about half of which are characterized by closed drainage (Dohrenwend 1987). Many of these closed basins contained pluvial lakes during the late Pleistocene (Mifflin and Wheat 1979). The focus of the EM Project is on small (less than 100 square kilometer) upland watersheds within the Shoshone, Toiyabe, Toquima, and Monitor Ranges. These catchments are underlain by a wide variety of rock types that reflect the complex structural geology of the area (Kleinhampl and Ziony 1985). Most of the upland watersheds terminate in large, well-developed alluvial fans along the range fronts.

Because of the large elevational gradients, precipitation and temperature varies significantly from the upper elevations to the mouths of the watersheds. Precipitation approaches 55 centimeters at upper elevations but decreases to as little as 20 centimeters in the central valleys. At Austin, Nevada (2,168 meters), where one of the few long-term weather stations in the area is located, annual average precipitation is 34 centimeters. Most precipitation falls during the winter as snow, and peak runoff and most flood events occur during snowmelt in late May to early June. Convective summer storms occasionally result in flash floods. The upland watersheds are typically characterized by narrow valleys, and the stream systems are high gradient, coarse-grained, and often highly incised (fig. 1.2). Although low flows in these stream systems range from about 0.015 to 0.063 cubic meters per second, high flows can range from 0.214 to 0.683 cubic meters per second (Hess and Bohman 1996). Precipitation and stream flows are highly variable both within and among years (fig. 1.3). Most of the change in stream channel pattern and form occurs during high-magnitude, low-frequency flows (Miller et al. 2001). Although many of the stream systems are ephemeral, the EM Project has focused primarily on perennial streams. Most of these streams can be categorized as second or third order at the point where they flow into the central valleys (as measured on 1:24,000 topographic maps).

Vegetation changes significantly along the elevational gradient from the central valleys to the mountain peaks. Valley floors are characterized primarily by salt desert vegetation, including *Sarcobatus vermiculatus* (greasewood), *Distichlis spicata stricta* (saltgrass), *Atriplex confertifolia* (shadscale), and *Kraschennikovia lanata* (winterfat) communities. On the broad alluvial fans of the central Great Basin, *Artemisia tridentata wyomingensis* (Wyoming big sagebrush) and *A. nova* (black sagebrush) communities dominate. At low- to mid-elevations within the upland watersheds, Wyoming big sagebrush communities intergrade with singleleaf pinyon (*Pinus monophylla*) and Utah juniper (*Juniperus osteosperma*) woodlands. At higher elevations, *Artemisia tridentata vaseyana* (mountain big sagebrush), *A. arbuscula* (low sagebrush), and mountain brush communities dominate. Occasional *P. flexilis* (limber pine) communities also are found at these higher elevations. Riparian vegetation consists of narrow stringers of *Populus tremuloides* (quaking aspen), *Betula occidentalis* (water birch),

FIGURE 1.2. (top) San Juan Creek basin—a typical upland drainage basin in central Nevada; (bottom) Typical, coarse-grained nature of the stream channels.

P. angustifolia and *P. trichocarpa* (cottonwoods), *Salix* spp. (willows), and *Rosa woodsii* (wild rose). Meadow communities occur in valley segments with elevated water tables, and are dominated by grasses, *Carex* spp. (sedges), and *Juncus* spp. (rushes). A detailed ecological type classification has been developed for central Great Basin riparian areas based on physical and hydrological characteristics, key soil attributes, and vegetation

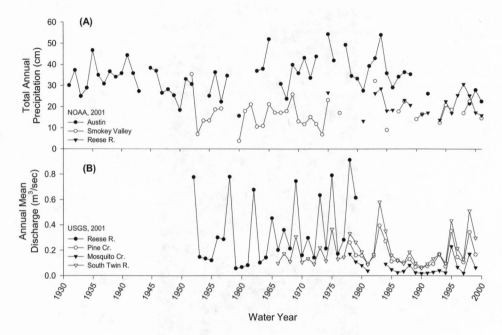

FIGURE 1.3. (A) The total annual precipitation (centimeters) measured for Smokey Valley, Reese River, and Austin, Nev.; and (B) the annual mean discharge (cubic meters per second) for selected stream systems in central Nevada.

species composition (Weixelman et al. 1996). A broader-scale classification of Great Basin riparian areas also exists (Manning and Padgett 1995).

In the state of Nevada, the federal government has jurisdiction over about 87 percent of the land area. The Bureau of Land Management manages 69 percent of the federal lands, while the United States Department of Agriculture (USDA), Forest Service, is responsible for 8 percent of these lands. The Forest Service has primary responsibility for management of the upland watersheds that are the focus of this book. At the base of the mountains, jurisdiction typically shifts from the USDA Forest Service to the United States Department of the Interior (USDI), Bureau of Land Management. There are small private inholdings in the upland watersheds, and large tracts of privately owned land in the broad central valleys. The majority of the perennial streams are diverted to private ranches in the central valleys at the base of the upland watersheds where they flow onto the alluvial fans.

Approach

There is a critical lack of information on watershed and stream processes as well as riparian ecosystem dynamics in the Great Basin. Clearly, no single approach is most appropriate for obtaining the information needed for management and restoration. Investigations must be designed to consider the physical and biotic characteristics of the system(s) of interest. The temporal and spatial scales over which information is gathered and analyzed are of primary importance because there is an inherent linkage among our scales of observation, the scales of the phenomena under consideration, and our ability to perceive change (Schumm and Lichty 1965; Frissel et al. 1986; Minshall 1988; Ritter et al. 2002). Moreover, the variables that have the greatest effects on the system vary with the scale at which the system is being observed. In extreme cases it is possible for changes in the scale of observation to result in a complete reversal of inferred cause-and-effect relationships. For example, over periods of years to decades, the slope of a channel is dependent on the hydrologic and sedimentologic regime of the stream (Schumm 1977; Knighton 1996; Ritter et al. 2002). Over shorter time spans, slope is relatively invariant and may be considered an independent variable that controls local flow and sediment transport conditions (Schumm 1977; Ritter et al. 2002). Thus, it is necessary both to account for the factors that determine the long-term dynamics of the fluvial system at the watershed scale and to document the factors that determine the dynamics of stream reaches and riparian ecosystems at smaller spatial and temporal scales. The following sections define the specific temporal and spatial scales of analyses that are used in the EM Project and that will be utilized throughout this volume.

Temporal Scales

The importance of historical information for understanding and predicting ecosystem change has long been recognized in the fields of geomorphology (Knox 1984) and hydrology (e.g., Poff 1991), and is increasingly recognized in ecology (Landres et al. 1999; Swetnam et al. 1999). Swetnam et al. (1999) suggest that the temporal scale examined should be sufficiently long to obtain meaningful information about species and populations, ecosystem and landscape structures, disturbance frequencies, process rates, trends, periodicities, and other system dynamics. In the semiarid Great Basin, the most spatially and temporally complete paleoecological and geomorphic records have been constructed for the mid- to late-

Holocene (Tausch and Nowak 2000; Miller et al. 2001). Because these records indicate that geomorphic responses to changing climates during the latter half of the Holocene significantly influence modern channel processes and riparian ecosystem dynamics (Miller et al. 2001), the EM Project has focused on this time period. Three important temporal scales have been examined: the mid-late Holocene, the post-settlement period, and the present.

For our purposes, the *mid-late Holocene* encompasses the last eight thousand years. Because different types of historical information have different degrees of resolution, they provide information applicable to different temporal scales. At the mid-late Holocene scale, pollen and packrat midden data provide information on changes in plant taxa and, consequently, on climate change. Stratigraphic and landform analyses provide information on the hydrogeomorphic changes that have occurred in response to climate change and tectonism.

The *post-settlement period* includes the time period since 1860. This period is examined separately from the mid-late Holocene because anthropogenic disturbances have had a significant influence on ecosystem processes in the Great Basin. For this time frame, stream gaging records (1950 to present) and atmospheric data (precipitation and temperature; 1930 to present) are used to reconstruct historical climate and flow regimes. Aerial photos (1955 to present) are used to obtain information on changes in stream channel pattern and form in response to climate and other natural and anthropogenic disturbances, and dendrochronology yields information on episodes of woody riparian species establishment, the geomorphic requirements (locations) of establishment, and the relationships of establishment to climate.

The *present* includes the time period of up to ten years before present and provides critical information on current ecosystem processes. At this scale, multiyear studies are used to examine the response of stream systems and riparian ecosystems to weather patterns, natural disturbances such as floods, anthropogenic disturbances such as roads and livestock grazing, and management or restoration treatments such as prescribed fire and revegetation.

Spatial Scales

Stream and riparian ecologists have recognized the utility of a hierarchical framework for understanding and categorizing watersheds and stream systems (Frissel et al. 1986; Minshall 1988; Maxwell et al. 1995). Hierarchical structuring (Allen and Starr 1982; O'Neill et al. 1986) means that, at a given level of resolution, a system is both composed of interacting components (lower-level entities) and is itself a component of a larger system (higher-level entity). Smaller-scale systems develop within constraints set by the larger-scale systems of which they are a part. Any given system is constrained by the potential behavior of its components and the environmental constraints imposed by higher levels. Watersheds, valley segments, and stream reaches are an example of hierarchical organization in stream systems (Frissel et al. 1986; Minshall 1988). The analysis conducted by the EM Project utilizes a four-part hierarchical system, each part defined by a unique spatial scale. The categorization of scales includes, from largest to smallest, watershed, riparian corridor, valley segment, and stream reach (fig. 1.4).

Historically, watersheds have been described as the most appropriate unit of study for ecosystem management and fluvial geomorphic analyses (Schumm 1977; Montgomery et al. 1995). Upland watersheds in the central Great Basin range in size from less than 5 to 100 square kilometers, are characterized by second- and third-order streams (as measured on 7.5′ topographic maps), are subjected to localized climatic conditions that vary with altitude over the basin, and exhibit a variety of vegetation types. At this scale, paleoecological reconstructions are used to determine temporal changes in vegetation and climate during the Holocene. These data are subsequently related to past alterations in geomorphic processes determined from alluvial stratigraphic and morphologic techniques. Ultimately, analyses at this scale are used to provide an understanding of the types of responses that result from environmental perturbations, and the sensitivity of watersheds to changing environmental conditions.

The riparian corridor is the integrated network of stream channels and adjacent geomorphic surfaces (i.e., floodplains and terraces) that primarily are located on alluvial deposits in the valley bottoms. The riparian corridor receives water and sediment from the surrounding hillslopes and represents a pivotal interface between surface and groundwater flow systems. Analyses at this spatial scale focus on the interactions between hillslope and channel processes and, through the use of remote sensing, vegetation types and patterns. These data are used to assess the relationships among landform assemblages, geomorphic processes, and vegetation.

Nested within the riparian corridors are valley segments that have semi-uniform valley characteristics (slopes, widths, geologic materials) and similar climatic conditions. At this scale, pack rat midden data are used to examine the effects of climate on local changes in riparian vegetation. The

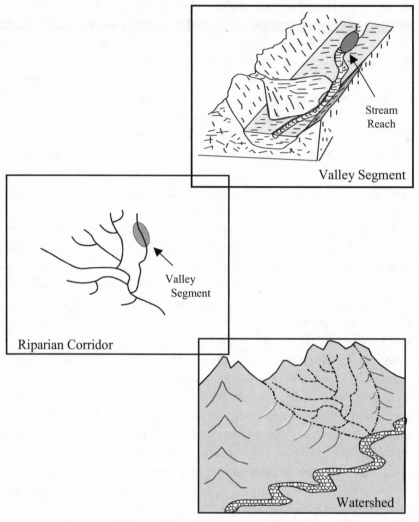

FIGURE 1.4. A schematic diagram of the hierarchical scales used in this volume: watershed, riparian corridor, valley segment, and stream reach.

individual and interrelated effects of landforms and stream incision are re-lated to riparian ecosystem dynamics. Also, reconstructions of climatic and hydrologic regimes in combination with dendrochronologic data are used to relate climate and stream flows to the establishment of woody riparian species.

The smallest units in our four-part hierarchy are stream reaches, defined as sections of the valley segments with relatively uniform channel mor-phology, bed material composition, bank conditions, and woody debris.

Linked geomorphic, hydrologic, and vegetation studies are again used, but the focus is on the kinds of ecosystem degradation that affect specific channel, vegetation, and soil types. The studies that are conducted are mechanistic in nature and are designed to increase our understanding of disturbance processes and potential restoration treatments.

Overview

The remaining chapters in this book provide integrated analyses of the data obtained from the geomorphic, hydrologic, and biotic investigations that have been conducted in the central Great Basin. The next seven chapters focus on specific areas of study and emphasize unique temporal and spatial scales. These chapters provide detailed overviews of the approaches and methods that have been used by the EM Project, summarize the findings of the various investigations, and discuss their implications. The final chapter attempts to integrate results of the individual investigations and to provide a more holistic understanding of the abiotic and biotic processes operating in riparian areas of the central Great Basin. It examines the relationships among these processes, the responses that may result from climate change and natural or anthropogenic disturbances, and the significance of these findings for ecosystem management and restoration.

The topics are generally arranged to take the reader from a discussion of large-scale, long-term analyses to shorter-term, more localized investigations. However, each chapter has been written as a self-contained document, making it possible to gain insights into each individual topic without necessarily reading the entire book.

The foundation for our discussions of the effects of climate change on the geomorphic and hydrologic processes and the riparian ecosystems of the central Great Basin are provided in chapter 2. This chapter gives an overview of climatic variation during the Holocene in the Northern Hemisphere, and examines the relationships between Holocene climate and vegetation dynamics in the Great Basin as revealed by paleoecological and paleoclimatic records. The authors suggest that variations in the diversity and relative composition of tree, shrub, grass, and forb taxa are strongly correlated with patterns of regional and hemispheric climate changes depicted in other paleoclimatic records based on patterns of ice rafting and deep-ocean cores. Eight climatic periods are derived for the Holocene that are characterized by unique vegetation associations. The changes that have occurred in vegetation patterns since European settlement of the Great Basin around 1860 also are described, and it is argued that changes in vegetation during the historical period result from complex interactions between climate change and anthropogenic factors. The correlations among vegetation and climate have important implications for future changes in Great Basin vegetation that could result from global warming, and the authors conclude with a discussion on the need for management approaches that take these potential changes into account.

The basis for stream and riparian ecosystem management is a sound understanding of fluvial geomorphic responses to both climate change and natural or anthropogenic disturbances. Chapter 3 expands on the discussion initiated in the previous chapter to include the geomorphic responses that resulted from changes in Holocene climate. The analysis is unique in that it compares the similarities and differences of responses to shifting climates at two different spatial scales. The geomorphic history of both the middle Humboldt River near Battle Mountain, Nevada, and of the small, upland watersheds of the Toiyabe, Toquima, and Monitor Ranges is synthesized. The authors reconstruct the erosional and depositional events that occurred in upland watersheds of the central Great Basin (the focus of most of the chapters in this volume) during the mid- to late-Holocene. They illustrate that climate changes that occurred over 2000 YBP (years before present) are still influencing modern geomorphic processes. Importantly, they present substantial evidence that the dominant response of stream channels to climate change and disturbance over the past 2,000 years has been incision, and that the most recent episode of channel incision began 400 to 500 YBP. Because the streams currently exhibit a tendency to incise, they are highly sensitive to anthropogenic disturbances. This chapter forms the basis for our understanding of current stream processes and riparian ecosystem dynamics in upland watersheds of the central Great Basin. It clearly demonstrates the need to consider past climate change and the resulting geomorphic responses as important controls of modern channel processes.

The understanding of geomorphic responses to climate change developed in chapter 3 serves as the basis for examining the sensitivity of stream systems in upland watersheds to natural or anthropogenic disturbance in chapter 4. The authors demonstrate that the sensitivity of watersheds to disturbance is a function of several controlling factors including geology, hydrologic regime, and basin morphometry (size, relief, etc.), and that the watersheds can be categorized on the basis of the rate and magnitude of recent channel incision. The basins fall into four well-defined groups on the basis of their response to environmental change, including Group 1, flood dominated, Group 2, deeply incised, Group 3, fan dominated, and Group 4, pseudostable. The controls on basin sensitivity to incision are complex—individual watersheds have unique responses. Nonetheless, the authors clearly show the advantages of documenting differences in regional watershed sensitivity to anthropogenic disturbances for effective management. The general approach is applicable elsewhere, although the categories developed use indicators that are specific to the geomorphic responses of upland watersheds in the central Great Basin to disturbance, and it would be necessary to develop indicators specific to the region of interest.

Both surface and subsurface water supplies are critical for sustaining riparian ecosystems. Riparian meadow complexes often are located in incision-dominated watersheds and are highly susceptible to channel incision. Chapter 5 focuses on the surface and groundwater flow systems supporting riparian meadow complexes within representative, upland drainages in the central Great Basin. The authors focus on the interactions between surface water and groundwater flow systems, the control(s) of groundwater regimes on vegetational communities, and the effects of channel incision on water table elevations and, thus, vegetational patterns. They show that the locations of meadow complexes reflect large-scale controls on basin shape and geology and that meadows are formed by upwelling groundwater that results from constrictions in the thickness of conductive valley-fill sediments. When incision occurs, the hydrologic patterns depend on the depth of entrenchment, and the distance downstream from the nearest constant head source supporting the meadow. The vegetation types within these meadow complexes are strongly correlated with water table depths and, for incised systems, the depth of incision. The authors suggest that effective management of these systems requires an understanding of the hydrological conditions leading to meadow formation, the base level controls on both streams and water tables, the susceptibility of soils and sediments to incision, and the sensitivity of the existing vegetation to changes in water tables.

An important ecosystem service provided by riparian areas is a supply of high-quality water that is adequate for both sustaining riparian ecosystems and supporting the needs of downstream users. Chapter 6 describes a large-scale survey of surface water quality conducted in the Monitor, Toquima, and Toiyabe Ranges in the spring of 1994, and a more detailed seasonal survey of water quality at multiple sites along streams in the Toiyabe Range conducted in 1995 through 1998. The authors use these water quality surveys to assess the relative importance of basin lithology, climate, and land-use and management activities as controls of water quality. In upland watersheds of the central Great Basin, basin lithology is the primary regulator of the chemical composition of streams, but streamflow influences stream chemical composition seasonally and annually (wet versus dry years). Stream chemical composition is not influenced by land-use activities in upland watersheds but is affected by agricultural activities in streams or rivers occurring in the broad central valleys. Because stream chemical composition is regulated by catchment lithology and streamflow (climate driven) in upland watersheds, this indicator of water quality may be sensitive to anthropogenic inputs in some basins should disturbance regimes or land-use activities change (e.g., increased fire frequency and severity, grazing or recreation intensity and duration).

Riparian ecosystems are influenced not only by geomorphic processes and hydrologic regimes but also by biotic interactions. In chapter 7, the relationships among riparian vegetation and geomorphic and hydrologic processes in upland watersheds of the central Great Basin are evaluated over a range of scales from a series of case studies. The authors show that the composition and pattern of streamside vegetation is highly correlated with the geologic and hydrologic characteristics of the watersheds and, thus, basin sensitivity to disturbance as categorized in chapter 4. They also demonstrate that the composition and pattern of the riparian vegetation is strongly influenced by the occurrence of side-valley alluvial fans within many of the watersheds. Meadow complexes often occur in incision-dominated basins, such as basins dominated by side-valley alluvial fans, are highly susceptible to stream incision, and are a management priority. The authors' studies show that riparian meadow complexes occur along hydrologic gradients, and that soil physical and chemical properties, plant physiological processes, and plant population and community dynamics are all closely related to the hydrologic regime as indicated by water table depths. Anthropogenic disturbances, such as overgrazing by livestock, can significantly alter plant interactions and community dynamics within these ecosystems. The authors argue that detailed knowledge of geomorphic processes and hydrologic regimes is necessary not only to understand the structure and composition of riparian vegetation, but also to predict vegetational responses to stream incision and to anthropogenic disturbances.

Understanding the effects of climate change and natural and anthropogenic disturbances on native faunal distributions is a critical aspect of maintaining and restoring native species and ecosystems. Chapter 8 presents an overview of four approaches to studying and managing faunal distributions in the Great Basin: documentation of historical changes; development of explanatory and predictive models; application of surrogate species, such as indicators or umbrellas, as planning tools; and use of island biogeographic theory to anticipate ecological effects of climate change. The authors illustrate both single- and multiple-species approaches using case studies from their work on native fishes, butterflies, and birds. The approaches used by the authors often have focused on links between target species and aspects of the physical environment, such as elevation or topographic heterogeneity. The authors effectively argue that an understanding of species responses to physical environmental variables can be a powerful and practical method for explaining why faunal distributions may have shifted in the past. These responses also can be used for predicting how those distributions may be affected by future environmental changes.

In the concluding chapter, knowledge from earlier chapters on the relationships among the geomorphic, hydrologic, and biotic processes structuring streams and riparian ecosystems is used to devise management and restoration strategies. Because many stream systems in the central Great Basin currently have a natural tendency to incise, they are functioning as nonequilibrium systems and restoring them to conditions that existed prior to incision is unrealistic. It is recommended that the management focus should be to restore and maintain stable stream and riparian ecosystems, and to slow the rates of change in incising systems. At the watershed–riparian corridor scale, a holistic approach can be used that is based on the basin sensitivity categories developed in chapter 4 and that includes management techniques such as proper grazing, stream stabilization, and road realignment. At the valley segment–stream reach scale, an understanding of the ecological types, alternative states, and thresholds that currently exist for incised or degraded ecosystems can be used to determine the site or restoration potential. Riparian ecosystems associated with incised stream systems or altered hydrologic regimes often have crossed an abiotic threshold and represent a new state with a site potential that differs from that prior to incision. The recognition of threshold crossings has led to an approach for defining the site potential and deciding upon restoration techniques that utilize state and transition models. The sensitivity of these watersheds to disturbance indicates that proactive management of both riparian ecosystems and their linked watersheds will be needed to prevent further degradation.

All too often, the benefits that could be derived from research endeavors are not passed along to those responsible for land management or for developing restoration initiatives. Inherent in the papers of this volume is an attempt both to further our scientific understanding of the abiotic and biotic processes structuring the riparian areas of the Great Basin and to apply the knowledge obtained to the development of sound management and restoration strategies. Clearly, there are many avenues of research that remain to be explored within the framework addressed by this volume. Nonetheless, we hope that the pages contained herein will not only summarize the current state of knowledge, but will stimulate and guide future research on the topic.

Acknowledgments

This chapter is the product of many discussions with the managers and researchers who are a part of the EM Project. We thank Julie Stromberg and Bob Beschta for thoughtful reviews of a draft manuscript.

Literature Cited

Allen, T. F. H., and T. B. Starr. 1982. *Hierarchy: Perspectives for ecological complexity.* Chicago: University of Chicago Press.

Belsky, A. J., A. Matzke, and S. Uselman. 1999. Survey of livestock influences on stream and riparian ecosystems in the western United States. *Journal of Soil and Water Conservation* 51:419–431.

Briggs, M. K. 1996. *Riparian ecosystem recovery in arid lands: Strategies and references.* Tucson: University of Arizona Press.

Chambers, J. C., K. Farleigh, R. J. Tausch, J. R. Miller, D. Germanoski, D. Martin, and C. Nowak. 1998. Understanding long- and short-term changes in vegetation and geomorphic processes: The key to riparian restoration. Pp. 101–110 in *Rangeland management and water resources,* edited by D. F. Potts. Herndon, Va.: American Water Resources Association and Society for Range Management.

Chapin, F. S. III., M. S. Torn, and M. Tateno. 1996. Principles of ecosystem sustainability. *American Naturalist* 148:1016–1037.

Christensen, N. L., A. M. Bartuska, J. H. Brown, S. Carpenter, C. D'Antonio, R. Francis, J. F. Franklin, J. A. MacMahon, R. F. Noss, D. J. Parsons, C. H. Peterson, M. G. Turner, and R. G. Woodmansee. 1996. The report of the Ecological Society of America Committee on the scientific basis for ecosystem management. *Ecological Applications* 6:665–691.

Dahm, C. N., and M. C. Molles Jr. 1991. Streams in semiarid regions as sensitive indicators of global climate change. Pp. 250–260 in *Climate change and freshwater ecosystems,* edited by P. Firth and S. G. Fisher. New York: Springer-Verlag.

Dohrenwend, J. C. 1987. Basin and range. Pp. 303–342 in *Geomorphic systems of North America,* edited by W. L. Graff. Centennial special volume 2. Boulder, Colo.: Geological Society of America.

Dunham, J. B., G. L. Vinyard, and B. E. Rieman. 1997. Habitat fragmentation and extinction risk of Lahontan cutthroat trout (*Oncorhynchus clarki henshawi*). *North American Journal of Fisheries Management* 17:910–917.

Elmore, W., and B. Kauffman. 1994. Riparian and watershed systems: Degradation and restoration. Pp. 212–231 in *Ecological implications of livestock herbivory in the West*, edited by M. Vavra, W. A. Laycock, and R. D. Pieper. Denver, Colo.: Society for Range Management.

Fleischner, T. L. 1994. Ecological costs of livestock grazing in western North America. *Conservation Biology* 8:629–644.

Fleishman, E., D. D. Murphy, and G. T. Austin. 1999. Butterflies of the Toquima Range, Nevada: Distribution, natural history, and comparison to the Toiyabe Range. *Great Basin Naturalist* 59:50–62.

Frissel, C. A., W. J. Liss, C. E. Warren, and M. D. Hurley. 1986. A hierarchical framework for stream habitat classification: Viewing streams in a watershed context. *Environmental Management* 10:199–214.

Goodwin, C. N., C. P. Hawkins, and J. L. Kershner. 1997. Riparian restoration in the western United States: Overview and perspective. *Restoration Ecology* 5:4–14.

Graf, W. L. 1988. *Fluvial processes in dryland rivers*. New York: Springer-Verlag.

Gregory, S. V., F. J. Swanson, W. A. McKee, and K. W. Cummins. 1991. An ecosystem perspective of riparian zones: Focus on links between land and water. *BioScience* 41:540–551.

Hess, G. W., and L. R. Bohman. 1996. *Techniques of estimating streamflow at gaged sites and monthly streamflow duration characteristics at ungaged sites in central Nevada*. Open-file report 96-559, U.S. Geological Survey. Carson City, Nev.

Hobbs, R. J., and J. A. Harris. 2001. Restoration ecology: Repairing the earth's ecosystems in the new millennium. *Restoration Ecology* 9:239–246.

Hobbs, R. J., and D. A. Norton. 1996. Commentary: Towards a conceptual framework for restoration ecology. *Restoration Ecology* 4:93–110.

Jenson, S. E., and W. S. Platts. 1990. Restoration of degraded riverine/riparian habitat in the Great Basin and Snake River Regions. Pp. 367–398 in *Wetland creation and restoration: The status of the science*, edited by J. A. Kusler and M. E. Kentula. Covelo, Calif.: Island Press.

Kauffman, J. B., R. L. Beschta, N. Otting, and D. Lytjen. 1997. An ecological perspective of riparian and stream restoration in the western United States. *Fisheries* 22:12–24.

Kauffman, J. B., and W. C. Krueger. 1984. Livestock impacts on riparian ecosystems and streamside management implications . . . a review. *Journal of Range Management* 37:430–438.

Kleinhampl, F. J., and J. I. Ziony. 1985. *Geology of northern Nye County, Nevada*. Nevada Bureau of Mines and Geology. Bulletin 99A. University of Nevada, Reno.

Knighton, D. 1996. *Fluvial forms and processes: A new perspective*. London: Arnold.

Knox, J. C. 1984. Fluvial responses to small scale climate change. Pp. 318–342 in *Fluvial responses to small scale climate changes: Developments and applications of geomorphology*, edited by J. E. Costa and P. J. Fleisher. Berlin: Springer-Verlag.

Kusler, J. A., and M. E. Kentula, eds. 1990. *Wetland creation and restoration: The status of the science*. Covelo, Calif.: Island Press.

Landres, P. B., P. Morgan, and F. J. Swanson. 1999. Overview of the use of natural variability concepts in managing ecological systems. *Ecological Applications* 9:1179–1188.

Manning, M. E., and W. G. Padgett. 1995. *Riparian community type classification for the Humboldt and Toiyabe national forests, Nevada and eastern California*. R4-Ecol-95-01. Ogden, Utah: USDA Forest Service, Intermountain Region.

Martin, T. E., and D. M. Finch. 1996. *Ecology and management of Neotropical migratory birds.* New York: Oxford University Press.

Maxwell, J. R., C. J. Edwards, M. E. Jensen, S. J. Paustian, H. Parrott, and D. M. Hill. 1995. *A hierarchical framework of aquatic ecological units in North America (Neartic Zone).* General Technical Report NC-176, USDA Forest Service, North Central Forest Experiment Station. St. Paul, Minn.

Mifflin, M. D., and M. M. Wheat. 1979. *Pluvial lakes and estimated pluvial climates of Nevada.* Nevada Bureau of Mines and Geology. Bulletin 94. Reno, Nev.

Miller, J., D. Germanoski, K. Waltman, R. Tausch, and J. Chambers. 2001. Influence of late Holocene hillslope processes and landforms on modern channel dynamics in upland watersheds of central Nevada. *Geomorphology* 38:373–391.

Minshall, G. W. 1988. Stream ecosystem theory: A global perspective. *Journal of the North American Benthological Society* 7:263–288.

Montgomery, D. R., G. E. Grant, and K. Sullivan. 1995. Watershed analysis as a framework for implementing ecosystem management. *Water Resources Bulletin* 31:369–386.

National Research Council. 1992. *Restoration of aquatic ecosystems: Science, technology, and public policy.* Washington, D.C.: National Academy Press.

——. 2002. *Riparian areas: Functions and strategies from management.* Washington, D.C.: National Academy Press.

Naveh, Z. 1994. From biodiversity to ecodiversity: A landscape-ecology approach to conservation and restoration. *Restoration Ecology* 10:58–62.

Ohmart, R. D. 1996. Historical and present impacts of livestock grazing on fish and wildlife resources in western riparian habitats. Pp. 245–279 in *Rangeland wildlife,* edited by P. R. Krausman. Denver, Colo.: Society for Range Management.

O'Neill, R. V., D. L. DeAngelis, J. B. Waide, and T. F. H. Allen. 1986. Hierarchical concept of ecosystems. *Monographs in Population Biology* 23:1–272.

Osterkamp, W. R., and J. M. Friedman. 2000. The disparity between extreme rainfall events and rare floods: With emphasis on the semi-arid American West. *Hydrological Processes* 14:2817–2829.

Overbay, J. C. 1992. Ecosystem management. Pp. 3–15 in *Taking an ecological approach to management.* Publication WO-WSA-3. Washington, D.C.: USDA Forest Service.

Poff, N. L. 1991. Regional hydrologic response to climate change: An ecological perspective. Pp. 88–115 in *Climate change and freshwater ecosystems,* edited by P. Firth and S. G. Fisher. New York: Springer-Verlag.

Ritter, D. F., R. C. Kochel, and J. R. Miller. 2002. *Process geomorphology.* 4th ed. Dubuque, Iowa: Wm. C. Brown Publishers.

Roper, B. B., J. J. Dose, and J. E. Williams. 1997. Stream restoration: Is fisheries biology enough? *Fisheries* 22:6–11.

Schumm, S. A. 1977. *The fluvial system.* New York: John Wiley & Sons.

Schumm, S. A., and R. W. Lichty. 1965. Time, space, and causality in geomorphology. *American Journal of Science* 263:110–119.

Sidle, R. C., and M. C. Amacher. 1990. Effects of mining, grazing, and roads on sediment and water chemistry in Birch Creek, Nevada. Pp. 463–472 in *Watershed planning and analysis in action.* Symposium Proceedings of the IR Conference on Watershed Management, American Society of Civil Engineers, Durango, Colo.

Sidle, R. C., and J. W. Hornbeck. 1991. Cumulative effects: A broader approach to water quality research. *Journal of Soil and Water Conservation* 46:268–271.

Society for Ecological Restoration International Science. 2002. *The SER primer on ecological restoration*. Science and Policy Working Group. http://www.seri.org.

Swetnam, T. W., C. D. Allen, and J. L. Betancourt. 1999. Applied historical ecology: Using the past to manage the future. *Ecological Applications* 9:1189–1206.

Tausch, R. J., and C. L. Nowak. 2000. Influences of Holocene climate and vegetation changes on present and future community dynamics. *Journal of Arid Land Studies* 10S:5–8.

Trimble, S. W., and A. C. Mendel. 1995. The cow as a geomorphic agent: A critical review. *Geomorphology* 13:233–253.

Wade, P. W., A. R. G. Large, and L. C. De Waal. 1998. Rehabilitation of degraded river habitat: An introduction. Pp. 1–12 in *Rehabilitation of rivers: Principles and implementation*, edited by L. C. Waal, A. R. G. Large, and P. W. Wade. Chichester, UK: John Wiley & Sons.

Weixelman, D. A., D. C. Zamudio, and K. A. Zamudio. 1996. *Central Nevada riparian field guide*, R6-ECOL-TP. Ogden, Utah: USDA Forest Service, Intermountain Region.

Whisenant, S. G. 1999. *Repairing damaged wildlands: A process-oriented, landscape-scale approach*. Cambridge, UK: Cambridge University Press.

Williams, J. E., C. A. Wood, and M. P. Dombeck. 1997. Understanding watershed-scale restoration. Pp. 1–16 in *Watershed restoration: Principles and practice*, edited by J. E. Williams, C. A. Wood, and M. P. Dombeck. Bethseda, Md.: American Fisheries Society.

Chapter 2

Climate Change and Associated Vegetation Dynamics during the Holocene: The Paleoecological Record

Robin J. Tausch, Cheryl L. Nowak, and
Scott A. Mensing

Dramatic changes have occurred in Great Basin plant communities over the last century, many of which are discussed in this volume. Coping with these changes requires an understanding of their dynamics and the causes of the many forces involved in driving them. As our knowledge of these systems has improved, it has become apparent that the most significant forces operate over long temporal and large spatial scales. Through its influence on energy balance and water availability, climate is one of the significant long-term factors driving changes in the occurrence and distribution of ecosystems and communities (Bailey et al. 1994). Landform morphology has a major spatial influence on climate, particularly in a region as topographically variable as the Great Basin. In combination, climate, geographic location, and topography have resulted in a complex mix of environmental conditions and vegetation communities in the Great Basin (Billings 1951).

Long-term changes in climate affect the key ecological processes that have driven the complex mix of both local and regional vegetation changes known to have occurred in the Great Basin (Betancourt et al. 1993; Woolfenden 1996; Wigand and Rhode 2002). These climatic changes, interacting with strong north-to-south and west-to-east gradients in community composition (Wigand and Rhode 2002), have dictated the timing and direction of community change throughout the Holocene (Thompson 1990). Today, these environmentally driven changes are interacting with anthropogenic-based impacts.

Considerable information is now available on climate changes and associated changes in vegetation that have occurred during the Holocene. The Holocene represents the most recent of several interglacials in long-term cycles between glacial and interglacial periods of the Pleistocene that

24

have occurred over the last 2.5 million years. The combination of paleo-botanical data from woodrat middens, pollen records, and tree rings in the Great Basin allow us to reconstruct vegetation and climate change across a wide range of spatial and temporal scales (Wigand and Rhode 2002). These can be correlated well with other records of global climate change (Benson et al. 1997) and demonstrate that vegetation changes over the Holocene are the result of the individualistic responses of plant species (Betancourt 1996; Betancourt et al. 1990; Nowak et al. 1994a; Tausch et al. 1993; Van Devender and Spaulding 1979; Woolfenden 1996). Throughout the Holocene, plant communities have been loosely knit assemblages that continually shift in composition through time (Tausch et al. 1993; Wigand and Rhode 2002) in response to changing climate (Whitlock and Bartlein 1997).

From the information now available for the Great Basin, it is evident that the influence of past environmental conditions can often equal or exceed the importance of current environmental conditions in determining community dynamics (Millar 1996; Woolfenden 1996; Tausch 1999). Thus, a knowledge of long-term change is necessary to fully understand current community patterns or their ongoing changes (Tausch 1996). The changes that have occurred in central Nevada riparian communities during the Holocene described in chapter 7 of this volume support this view.

The effect of climate on ecosystems involves not only changes in averages, but also changes in extremes. Changes in the magnitude and timing of climatic extremes are particularly important because they often have the most influence on community dynamics (Betancourt et al. 1993). Changes in the seasonality of precipitation as influenced by the relative importance of summer monsoonal precipitation versus that from winter Pacific frontal systems (Pyke 1972) have also resulted in regional differences among Great Basin communities (Wigand and Rhode 2002) that influence their response to climate change. Vegetation response to the variations in climate is thus complex, reticulate, and cumulative. It is dependent upon the plant species involved, the communities in which they are found, and geographic location. The sensitivity of vegetation communities to climate change is a summation of the physiological tolerances, adaptations, and life-history characteristics of the individual species present (Wigand and Rhode 2002; Stutz 1984), and chance events.

Until recently, responses to past climatic variations have only rarely been studied at the ecosystem or regional scale (Betancourt et al. 1993).

Increasingly, paleostudies are focusing on regional interpretation, and scientists are gaining a better understanding of the dynamics of community responses at ecosystem and regional scales that is critically needed to effectively manage riparian and upland ecosystems in the Great Basin. Better management requires the incorporation of such an expanded perception of plant and community changes to time scales more appropriate for viewing long-term responses to climate change (Millar 1997; Tausch 1996, 1999; Tausch and Nowak 2000). The studies reported in this chapter, as well as in other chapters within this volume, help to address these needs.

An important source of information regarding long-term plant responses is that obtained through paleoecological studies, which independently yield information on vegetation change and associated climate change over time periods from centuries to millennia. Paleoclimatic and paleovegetation data are in turn essential to assessing past changes in geomorphic-hydrologic systems. Our focus in this review is on the last 10,500 YBP (years before present; the Holocene). In recent decades, studies of macrofossils from woodrat middens, fossil pollen, tree rings from live and dead wood, glacial ice cores, marine and pluvial lake sediment cores, and variation in glacial activity (Denton and Karlen 1973; Thompson et al. 1993; Tausch 1999; Wigand and Rhode 2002) have increased our ability to identify periods of climatic constancy and variation and corresponding vegetative response. As this information has been accumulating, periodic attempts have been made to delineate these patterns into distinct climatic phases through the Holocene (Tausch and Nowak 2000; Madsen et al. 2001; Benson et al. 2002; Wigand and Rhode 2002). These delineations have generally been broad and subjectively determined by looking for corresponding patterns of variation in results from various sources of paleoinformation.

The various attempts to produce climatic chronologies for the Holocene have resulted in subdivisions with both similarities (Wigand and Rhode 2002) and differences between them. There may be a number of reasons for the differences between chronologies. One important aspect is that differences, or contrasts, can exist in the regional response to climate change, resulting in somewhat different vegetation responses to the same climate-forcing function. There can also be a transgressive nature to change across a continent in the effects of a climate-forcing function (Grayson 1993). Thus, a crucial element is to document where the patterns of change are similar in timing across the available data. The differences in how the various proxy data respond to climate changes also need to be considered.

There are also confounding factors in the interpretation of the various proxy data. Each specific proxy method may respond to only specific aspects of climate change. The responses that are present for such data may also be limited for some types of climate change. The more sources of proxy data that are included, the better the potential resolution of the effects of climate changes. Problems can also be due to limitations in the temporal resolution of the various data sets. For example, data points may be missing from key sections of the chronology. In addition, there are often large confidence limits and other problems with radiocarbon dates that reduce timing accuracy (Rhode 2000; Walker 2001). Identification of the timing of change can be further confounded by the possibility of climate and vegetation carryover influences, or inertia (Cole 1985; Lewin 1985; Davis 1986), associated with delayed relationships between the proxy data and the patterns of variation in the climate parameters of interest. Fortunately, the information available for the last five thousand years has increased sufficiently in recent years to provide a finer temporal resolution that can be used for a more detailed chronology of climate and associated vegetation change.

A key aspect of identifying important climatic periods (phases) during the Holocene is the individualistic nature of responses by plant species. Our ability to interpret responses resulting from climate variation is often hindered by the limited taxonomic resolution in the available data. In addition, each species has a unique tolerance range to environmental conditions. This range can be rather narrow for some species and very broad for others. One example can be seen in the presence or absence of sagebrush in the fossil record. Today, sagebrush steppe in the Great Basin occurs from the valley floors to the tops of some of the highest mountains. This plant community type encompasses a range of environmental conditions potentially equal to what has occurred for any single location across the entire Holocene. A variety of sagebrush species and subspecies, each with their own environmental tolerances occur over this entire elevational gradient (Winward 1969; West et al. 1978, 1998). However, very few of these taxa can currently be individually identified in paleovegetation data. Other species have narrower tolerances to climate and, thus, are more sensitive to change. Limitations in plant species distributions also can vary over space and time in ways not always directly related to climate and can involve many kinds of community interactions (Wigand and Rhode 2002). These constraints should be considered when using paleovegetation data to identify major shifts in climate.

Reference Chronology Development

An understanding of global changes in climate that occurred during the Holocene, and identification of the possible mechanisms driving these changes, has improved considerably in recent years by the inclusion of additional proxy indicators. One of the best of these new indicators of variation is the recent documentation of millennial-scale quasi cycles of ice rafting from glaciers into the North Atlantic Ocean, reconstructed from fine-grained terrestrial materials recovered in ocean cores (Bond et al. 1997, 2001; Broecker 2001). These cycles have been shown to be closely correlated with cycles in the production rate of the isotopes carbon-14 and beryllium-10 in the atmosphere (Bond et al. 2001; Kerr 2001). These two isotopes are produced by cosmic rays striking the upper atmosphere and are then deposited in tree rings and glacial ice, respectively (Bond et al. 2001). Variations in the production of these isotopes during the Holocene are the result of variation in the solar energy reaching the earth's atmosphere. When the sun emits more energy, increased solar winds reduce the impact of cosmic rays on the earth's atmosphere, and hence reduce the production of isotopes (Bond et al. 2001).

The small differences in the sun's energy indicated by the changes in isotope levels, however, are not sufficient to directly drive the observed changes in climate. Differences in insolation are apparently triggering other atmospheric changes that drive climate changes which are ultimately reflected in the variation in the quantity of drift ice in the North Atlantic (Bond et al. 2001). Fluctuations in the levels of Lake Bonneville over the last ten thousand years of the Pleistocene (21,000 to 11,000 YBP) are also apparently correlated with variations in the ice rafting into the North Atlantic (Oviatt 1997). Similar cycles have been observed for Summer Lake in the northwestern Great Basin (Zic et al. 2002). Stine (1990b) also noted a possible connection between variations in the sun's energy and long-term variation in Great Basin climate.

The patterns of ice rafting documented by Bond et al. (2001) provide some of the best information currently available concerning the large-scale patterns of climatic variation in the Northern Hemisphere throughout the Holocene. The illustration from Bond et al. (2001) used here (fig. 2.1) represents an averaging (stacking) of four records, two each from the eastern and western Atlantic Ocean, that demonstrate the regional drift-ice variability. Each record provides data on percentage variations in the quantity of specific petrologic tracers (identifiable terrestrial materials) in ocean

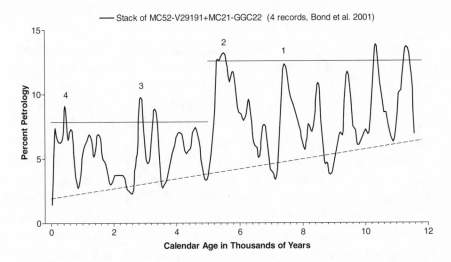

Figure 2.1. Diagram of average percent petrology adapted from figure 2 in Bond et al. (2001). This record represents an averaging (stacking) of the data for specific petrologic tracers from four oceanic cores from the North Atlantic that demonstrate variability in ice-rafting. The petrologic tracers are fine-grained terrestrial materials carried into the North Atlantic by the ice rafting. The two horizontal lines are the average of the peaks in percent petrology for before and after 5000 YBP. The dashed line is from a regression analysis of the minima in percent petrology by calendar year across the Holocene. Arrows indicate the boundaries between Great Basin climatic periods discussed in the text. The numbers indicate the four periods with the highest levels of drift ice over the last 8000 YBP.

Original figure reprinted (abstracted/excerpted) with permission from G. Bond, B. Kromer, J. Beer, R. Muscheler, M. N. Evans, W. Showers, S. Hoffman, R. Lotti-Bond, I. Hajdas, and G. Bonane. 2001. "Persistent solar influence on North American climate during the Holocene," Figure 2, *Science* 294:2130–2136, copyright 2001, The American Association for the Advancement of Science.

cores that reflect the change in the abundance of drift ice. The average of the individual petrologic tracers (fig. 2.1) is expressed as percent petrology (Bond et al. 2001). Ice rafting of terrestrial materials into the North Atlantic is much reduced during warm periods of higher solar activity, resulting in a lower percent petrology, but increases during periods of lower solar activity, resulting in peaks in percent petrology (Bond et al. 2001). A correlation analysis between the minima in average percent petrology and calendar year over the Holocene (fig. 2.1) was significant ($r = 0.85$). An average decline of about 0.38 percent for each one thousand years from 10,500 YBP to the present appears to represent a steady increase in temperature for the warmer periods of the Holocene.

Holocene Patterns of Vegetation Change

The patterns of variation in ice rafting (Bond et al. 2001) were used here as an independent reference (Walker 2001) of millennial scale cycles that cover the entire Holocene to help in identifying different periods of distinct climate in the Great Basin. We used two sources of variability apparent in the patterns of change in average percent petrology from Bond et al. (2001). These were (1) timing of the largest amplitude in variation in the drift-ice record, evaluated as the greatest percentage change in the shortest period of time, and (2) variation in the duration of drift-ice excursions, measured in years between minima or maxima in the petrology record (fig. 2.1). At specific points in time, proxy data from the Great Basin also identify environmental changes that point to significant shifts in climate. The timing of these changes generally coincides with the timing of large-amplitude variations in the production of ice-rafted debris, which correlate with changes in solar output.

We placed the boundaries between different Great Basin climatic periods at times when large-amplitude changes in the North Atlantic record correspond with large-magnitude landscape change in the Great Basin. This process results in identified boundaries that reflect the points in time when climate change was probably greatest (fig. 2.1). Because there has been a great deal of climatic variability within the Holocene, the proxy data also exhibit variability within the defined periods, and these periods should not be considered as times of unchanging climate. Rather, the periods represent specific times when the climate change was most dramatic and the associated evidence for vegetative change was the clearest. Our description of these periods covers the latest interglacial, the Holocene, starting after the Younger Dryas and covering the last 10,500 YBP.

Early Holocene: 10,500 to 7500 YBP

The early Holocene period described here incorporates the last four of five, relatively uniform millennial scale cycles of ice rafting in the North Atlantic Ocean (fig. 2.1). Climate during this early Holocene period was largely the continuation of the last stages of climate warming associated with the transition of the Pleistocene into the Holocene. Information from western Great Basin lakes (Benson et al. 2002) and the Bonneville Basin (Madsen et al. 2001; Wigand and Rhode 2002) indicates that an overall late Pleistocene drying trend occurred for about two millennia prior to 10,500 YBP (ages are given as calibrated radiocarbon years), followed by relatively wet, but variable conditions, apparently reflected in the relatively uniform periodicity in the cycling in ice rafting.

Associated with warming in the early Holocene was a general upward elevational shift in many species, a regional decline in the abundance of pine (*Pinus* spp.) and sagebrush (*Artemisia* spp.) and major increases in greasewood (*Sarcobatus* spp.) and saltbush (*Atriplex* spp.) in the western Great Basin (Thompson 1990; Wells 1983; Wigand et al. 1995; Wigand and Rhode 2002). Substantial range shifts occurred for a number of plant species in the eastern Great Basin (Wigand and Rhode 2002). Vegetation changes included the movement of pinyon-juniper woodlands from low-elevation slopes and valley floors of southern Nevada upward to mid-elevations, where they replaced the limber pine that were moving up to isolated mountain tops (Thompson 1990). The northward movement of pinyon and the expansion of juniper out of its more northerly refugia also began (Nowak et al. 1994a,b; Woolfenden 1996). Pinyon appears to have been a minor component of the woodlands in the southern Great Basin (Spaulding 1985, 1990; Thompson 1990; Wigand et al. 1995).

As the vegetation that is currently typical of the Great Basin moved to higher elevation and northward, plant species common to the Mojave Desert replaced them at the lower elevations (Mehringer 1977; Spaulding 1985, 1990). Climate in the southern Great Basin varied during this period as indicated by upward and downward shifts in the elevation of juniper (Wigand and Rhode 2002) while occurrences of flooding and desiccation alternated on some playas (Brown et al. 1990). The ice-rafting minimum at 8800 YBP (fig. 2.1) was associated with a conversion from sagebrush to desert shrub in valleys of the northwestern Great Basin (Wigand and Rhode 2002). It was also associated with the end of lithoid tufa deposition in Pyramid Lake (Benson et al. 1995). Species composition of all communities showed a general directional change reflecting more arid conditions during this period (Nowak et al. 1994a,b; Mensing 2001).

Mid-Holocene: 7500 to 5000 YBP

Following a peak at about 7500 YBP (fig. 2.1), ice rafting rapidly declined to a low of about 4 percent at approximately 7200 YBP and then did not return to its next major high point until about 5500 YBP. This interval between peaks in ice rafting of nearly two thousand years is the longest of the Holocene record. Many sources of evidence indicate that this long interval was the warmest, and possibly the driest, period of the Holocene that had occurred thus far for the Great Basin (Mehringer 1977, 1985; Wigand and Rhode 2002). Summer precipitation in the southern Great Basin appears to have been limited at this time (Spaulding 1991), but an increase may have occurred in the eastern Great Basin (Thompson 1984, 1990) accompanied by an increase in juniper and pine in the uplands (Thompson and Kautz 1983).

At various times during this period, upper tree lines, and vegetation communities in general, were as much as 300 to 500 meters higher in elevation than today (LaMarche 1974; Jennings and Elliot-Fisk 1993; Wigand et al. 1995; Woolfenden 1996). Lake Tahoe was as much as 4 or more meters below its natural rim between 7500 and 6300 YBP, and again about 1 to 1.5 meters below its rim around 5000 YBP, both coinciding with major declines in the abundance of drift ice (fig. 2.1). This drop in lake level has been determined from dating preserved trunks of trees that established during the lower lake levels and then drowned when the lake level rose again (Furgurson and Mobley 1992; Lindstrom 1990). Oxygen isotope and magnetic susceptibility values in Pyramid Lake indicate that Lake Tahoe was below its rim during much of this period (Benson et al. 2002), and Pyramid Lake may have reached its lowest levels (Born 1972). Walker Lake became a dry playa (Bradbury et al. 1989).

Many of the desert shrub species in the Great Basin experienced substantial increases in abundance during this period (Mehringer and Wigand 1990; Wigand 1987; Wigand et al. 1995; Wigand and Rhode 2002). In response to reduced vegetation cover, eolian processes increased, particularly in areas exposed by declining lake levels (Wigand and Mehringer 1985; Mehringer and Wigand 1990). Some expansion of the pinyon-juniper woodlands across the Great Basin may have occurred (Tausch 1999; Miller and Tausch 2001).

Post-Mid-Holocene Transition: 5000 to 3500 YBP

After 5000 YBP, the average height of peaks in percent petrology from ice rafting shifted from between 12 to 13 percent petrology prior to 5000 YBP down to an average of about 8 percent for the rest of the Holocene. The reasons for this drop are unknown, but it can be speculated that carryover of glacial ice from the Pleistocene into the early Holocene may be involved. The highest peaks in the post mid-Holocene transition were less than 8 percent petrology. This period is less well documented for the Great Basin than some others but seems to have generally been warm and dry with limited and erratic periods of increased precipitation and decreased temperature, in contrast to the mid-Holocene (Chatters and Hoover 1992;

Davis 1982; Mehringer 1987; Wigand 1987; Tausch and Nowak 2000; Wigand and Rhode 2002). The ^{18}O isotope record for Pyramid Lake also shows significant drought during this period (Benson et al. 2002), consistent with this long period of warmer and drier conditions. Many western Great Basin lakes remained low or desiccated during this period (Wigand and Rhode 2002) and Lake Tahoe was apparently below its rim for parts of this period (Benson et al. 2002). A slow migration of pinyon and juniper into and across the Great Basin generally continued at higher elevations. The first evidence of western juniper in northeastern California (Mehringer and Wigand 1990; Miller and Wigand 1994; Wigand et al. 1995; Wigand and Rhode 2002), Utah juniper in southern Montana (Lyford et al. 2002), and pinyon in the eastern Great Basin (Wells 1983) occurs at this time.

Neoglacial: 3500 to 2600 YBP

At about 3500 YBP, a rapid shift from minimum to maximum percent petrology occurred, with a peak of nearly 10 percent. This Neoglacial period is the first of the two periods that equaled or exceeded 9 percent petrology in the last half of the Holocene, and this period is clearly evident in many paleorecords. The Neoglacial was much cooler and wetter than the previous mid-Holocene interval or the transition that followed (Davis 1982; Grayson 1993; Wigand 1987; Woolfenden 1996; Tausch and Nowak 2000; Wigand and Rhode 2002). The increases in precipitation occurred almost entirely in the winter (Davis 1982; Wigand 1987). In the Bonneville Basin, this period saw one of the most dramatic cooling events of the Holocene (Madsen et al. 2001). In both the southern (Enzel et al. 1992) and the northern Great Basin (Wigand 1987; Madsen et al. 2001), evidence indicates that the Neoglacial may have been wetter than at any other time since the early Holocene (Wigand and Rhode 2002). Effects were more evident in the north and central than in the southern Great Basin (Wigand and Rhode 2002). A corresponding expansion of mountain glaciers has been reported for many locations in the Northern Hemisphere (Denton and Karlen 1973; Konrad and Clark 1998).

The expansion of pinyon-juniper woodlands in the central and southern Great Basin and of western juniper in the northwest accelerated during this period (Madsen 1985; Wigand 1987; Wigand and Rhode 2002). Most of the Holocene expansion of the woodlands at mid- and low elevations in the Great Basin had occurred by the end of this period (Mehringer and Wigand 1990; Wigand et al. 1995). The extent and density of the woodlands was possibly equal to that present today by the end of the period (Miller and Wigand 1994; Kinney 1996; Wigand et al. 1995). Desert shrub abundance decreased and grass and sagebrush increased during this time (Wigand 1987; Mehringer and Wigand 1990; Wigand and Rhode 2002). The elevation of the upper tree line decreased in the White Mountains (LaMarche 1973), in the Sierra Nevada (Scuderi 1987) and in the Canadian Rockies (Luckman 1990). Cool and wet conditions were also reflected in rising lake levels. The Great Salt Lake apparently expanded sufficiently during the latter part of the period to reach the Nevada-Utah border (Mehringer 1977; Madsen et al. 2001), Lake Tahoe was apparently spilling continuously, and Pyramid Lake rose (Benson et al 2002). Deep lake levels occurred in Walker Lake (Bradbury et al. 1989), and Silver Lake in the southern Great Basin had a high water level (Enzel et al. 1989). Mono Lake reached its highest level since the mid-Holocene around this time (Stine 1990a).

Post-Neoglacial Drought: 2600 to 1600 YBP

Between 2800 and 2500 YBP the percent petrology from the North Atlantic cores declined precipitously and remained low for about one thousand years (fig. 2.1). This was the third time in the Holocene that a major high in drift ice was followed by a new Holocene low. This post-Neoglacial drought is the longest period of the Holocene with minimal ice rafting—up to three times the length of any other minima. The percent petrology maxima on either side of this low period are separated by about 1,300 years, the second-longest interval between peaks in ice rafting in the Holocene record. The percent petrology record is matched by a period of extended drought in the Great Basin, with evidence for a significant drop in precipitation and lowered lake levels (Bradbury et al. 1989; Madsen et al. 2001; Wigand and Rhode 2002). Climatic variability was greater in the south than in the north (Mehringer and Warren 1976). Recent information (Miller et al. 2001; chapter 3) has shown that the dramatic decline in the precipitation of this period, and the associated decline in vegetation cover, was accompanied by a significant increase in hillslope erosion, the growth of alluvial fans, and sediment deposition in channel and valley bottoms over most of the upland drainages in the central Great Basin. Similar major geomorphic changes, including alluvial fan aggradations, also occurred in the eastern Columbia River Basin during this period (Chatters and Hoover 1992). Data from woodrat middens located adjacent to riparian areas in the Toiyabe Mountains (currently under analysis) show a drop in the number of plant taxa found in the middens during this period, from a high level during the Neoglacial, to the lowest levels of the Holocene (Chambers et al. 1998; Miller et al. 2001; R. J. Tausch and C. L. Nowak, unpublished data). Pollen evidence from Pyramid Lake and a wet-meadow pollen core in the Toiyabe Mountains (S. Mensing, unpublished data) also indicate a major drought period. Upper-elevation tree-ring evidence for this period for the region is limited (Anderson et al. 1997).

A decrease in woodlands and an increase in desert shrub vegetation dominated by Chenopods such as greasewood (*Sarcobatus vermiculatus*) occurred in the Great Basin during the post-Neoglacial drought (Thompson and Kautz 1983; Wigand 1987; Wigand et al. 1995; Wigand and Rhode 2002). Utah juniper, however, expanded its range north into the Wind River Canyon region of Wyoming (Jackson et al. 2002). Coinciding with the beginning of this period were additional declines in upper tree line for some locations in the southern Sierra Nevada (Scuderi 1987; Lloyd and Graumlich 1997) as well as some locations in the White Mountains (LaMarche 1973). Although this period was drier, precise temperatures are unknown, but they may have remained relatively cool (Chatters and Hoover 1992). By geographic contrast, moisture along the Texas Gulf Coast increased at this time (Hall 1990).

Post-Drought Transition and Medieval Warm Period: 1600 to 650 YBP

Ice rafting increased between 1600 YBP and about 1200 YBP, and then again declined precipitously, reaching a minimum at about 800 YBP. These three hundred to four hundred years of drier and cooler climate at the end of the period have sometimes been treated as a separate climate period (Tausch 1999; Tausch and Nowak 2000). Reconstruction of stream flow for the Sacramento River based on tree-ring records in the northern Sierra Nevada (Meko et al. 2001) found that one of the driest periods in the last one thousand years occurred at about 800 YBP. Although caution is needed in comparing Sierra Nevada records with those from the Great Basin, tree-ring records from the northern Sierra Nevada (Holmes et al. 1986; Woolfenden 1996) identify droughts at about 1150 YBP and 800 YBP. Growth patterns on now-submerged stumps from Lake Tahoe, Mono Lake, and Fallen Leaf Lake generally agree with these findings (Wigand and Rhode 2002). Oxygen isotope data from Pyramid Lake support the interpretation of extended droughts during this period (Benson et al. 2002). There was probably a significant carryover of the vegetation that was a consequence of the previous drought. This was due to both the limited increase and the variability in precipitation, in combination with the length of the drought that preceded this period. The abundance of pinyon-juniper woodlands also may have declined, although northward expansion continued (Wigand 1987; Wigand and Rose 1990). An increase in desert shrubs also occurred (Wigand et al. 1995). On the western Sierra Nevada there was an increase in fire frequency at this time (Swetnam 1993). Although more moisture was apparently available during this period compared to the previous period, overall it was still relatively low and variable.

The proxy data in the literature from the Great Basin region that is identified as the Medieval Warm Period, or Medieval Climate Anomaly, are inconsistent in that they are scattered over most of the period. For this reason the discussions that follow cover the entire period from 1600 YBP to 700 YBP. This period was apparently characterized by relatively warm temperatures (Grove and Switsur 1994), and an upward movement of tree lines has been documented between 1400 and 1000 YBP (Hiller et al. 2001). Near the middle of the period, a shift in the seasonality of precipitation occurred so that a greater proportion came in late spring or early summer (Wigand and Nowak 1992; Davis 1994; Leavitt 1994). This precipitation shift may have resulted in milder winter conditions (Wigand et al. 1995) and a reduction in snow pack, leading to reduced stream flows and lake levels (Born 1972; Stine 1994; Woolfenden 1996). The summer rain may have contributed to an expansion of pinyon in the southern Great Basin between 1600 and 1200 YBP (Wigand and Rhode 2002). Increasing summer rains resulted in an increase in grass abundance (Wigand 1987; Wigand and Nowak 1992) and a corresponding expanded presence of bison in both the eastern and western Great Basin (Schroedl 1973; Agenbroad 1978; Butler 1978; Wigand and Rhode 2002) for part of the period.

A maximum dominance of pinyon was centered at about 1200 YBP, coinciding with a peak in percent petrology (fig. 2.1). About this same time, the abundance of juniper briefly expanded in the north (Wigand et al. 1995). The Fremont Indian culture entered many areas of the eastern Great Basin shortly after the beginning of this period (Tausch 1999; Wigand and Rhode 2002) and were able to subsist on the available resources. The 800 YBP low point in the percent petrology (fig. 2.1) is reflected in some tree-ring studies (Holmes et al. 1986; Woolfenden 1996), but not in others (Esper et al. 2002; Briffa and Osborn 2002). Reduced lake levels (Stine 1990a), cool temperatures, and a decline in tree dominance (Wigand 1987; Wigand and Rose 1990) occurred along with an increase in desert shrubs and fire in some locations (Wigand et al. 1995). Prehistoric Indian cultures tended to move at this time from areas that had become inhospitable or failed to provide adequate resources. The Fremont Indian culture had disappeared from the Great Basin near the end of the period. In addition, the Anasazi culture in the Four Corners region began withdrawing about 820 YBP (Cordell and Gumerman 1989), and abandonment was completed by 650 YBP (Dean 1996).

Little Ice Age: 650 to 150 YBP

About 650 YBP, percent petrology (fig. 2.1) increased sharply to the last high period of the Holocene. The peak was approximately 9 percent midway through this period; the fourth major high point of the last 8000 YBP (fig. 2.1) and only the second time in the last half of the Holocene that this level was reached. As is evident in many paleorecords, the Little Ice Age had generally cooler and wetter conditions than the Medieval Climate Anomaly, particularly during the first half of this period (Woolfenden 1996). Conditions were similar to but more variable than the Neoglacial, with drier and more mild periods occurring. Glacial advances occurred, some of which were possibly the largest of the Holocene (Naftz et al. 1996; Clark and Gillespie 1997). Marshes and other wetlands expanded (Wigand and Rhode 2002), and lake levels increased in both the northern (Benson et al. 2002) and the southern Great Basin (Enzel et al. 1989). Growing season temperature remained low until the end of the period, and tree lines were possibly the lowest of the last seven thousand years, reflecting cold conditions (Stine 1996; Lloyd and Graumlich 1997).

Although both the Neoglacial and the Little Ice Age were relatively wetter and cooler, the record shows consistent differences between them. The Neoglacial was wetter than the Little Ice Age, and possibly less variable, as indicated by higher lake levels (Madsen et al. 2001) and larger meanders in the Humboldt River (chapter 3). However, the growth of mountain glaciers was generally greater during the Little Ice Age, again reflecting cold conditions (Clark and Gillespie 1997; Konrad and Clark 1998). A better understanding of the reasons for these differences is needed.

Pinyon-juniper woodlands continued a gradual increase in abundance and range during this period, recovering from the lower levels of abundance that had resulted from the post-Neoglacial drought (Mehringer and Wigand 1990; Wigand and Nowak 1992). These woodlands were dominated by western juniper in the northwest and primarily by pinyon over the rest of the Great Basin (Nowak et al. 1994a,b). This expansion was underway when the first Europeans arrived in the region (Miller and Wigand

1994; Wigand et al. 1995). The vegetation that was present when the Europeans arrived had developed during the previous five hundred years of the Little Ice Age (Woolfenden 1996; Tausch 1999). Current studies of woodrat middens in the Toiyabe Mountains (R. J. Tausch and C. L. Nowak, unpublished data) show that Great Basin riparian communities had as many plant taxa, particularly for the herbaceous species, as in any of the other wetter periods of the Holocene. Even though our understanding of Little Ice Age vegetation is better than for any prior period of the Holocene, major knowledge gaps persist because of limited temporal resolution in the available data.

During the Little Ice Age, the distribution and dominance patterns of pinyon-juniper woodlands and associated sagebrush ecosystems were very different than what is present today. Vegetation during the Little Ice Age seems to have been maintained by factors such as a cold, somewhat dry (compared to the Neoglacial) climate (Woolfenden 1996) and a higher fire frequency (Gruell 1999). Little Ice Age woodlands were more spatially variable, with trees located either in somewhat open savannas or in areas of woodland with higher densities that were confined to scattered fire-protected sites (Miller and Wigand 1984; Wigand et al. 1995).

Recent: 150 YBP to the Present

This period represents the final decline in percent petrology (fig. 2.1), to the lowest level of the Holocene (less than 2 percent). Pollen records in the Great Basin indicate increasing aridity and a reduction in grass abundance after the end of the Little Ice Age (Wigand 1987). This is also reflected in the rapid retreat in mountain glaciers around the world (Bradley 2000). In addition to climate change, many anthropogenically driven changes are occurring. These include (1) cessation of the Little Ice Age activities of hunting, gathering, and burning by populations of indigenous people (Creque 1996); (2) the period of heaviest livestock use in the region and associated effects on plant competition (Miller and Tausch 2001); (3) decreased fire frequency in native plant communities from both a removal of fine fuels by grazing and fire suppression efforts (Bunting 1994); (4) changing community composition resulting from altered competitive interactions associated with increasing CO_2 levels (Farquhar 1997; Smith et al. 2000); and (5) the increasing availability of nitrogen from air pollution (Kaiser 2001). The expansion of woodlands has been accelerated during this period (Miller and Tausch 2001). More recently, a substantial expansion in the presence and dominance of exotic species occurred in the Great

Basin (D'Antonio and Vitousek 1992). Because of these impacts and the changes they are driving, the dynamics and complexity of communities will likely never return to those of the past.

Management and Restoration Implications

The link between past changes in Great Basin vegetation and the patterns of ice rafting (fig. 2.1) provides a means of predicting future conditions in the area and their effect on restoration efforts. There are always problems in making projections, such as an unanticipated future occurrence of a threshold climate change, like the drop in the peaks of ice rafting after 5000 YBP. However, potential projections may still be useful in assessing future requirements for restoration. The four highest points in ice rafting over the last eight thousand years were each followed shortly afterward by a rapid decline to a drift-ice minima. Each minima represents the lowest level of Holocene drift ice up to that point in time (fig. 2.1). The last of those peaks in ice rafting, the Little Ice Age, was again followed by a rapid decline to what is already the lowest level of drift ice of the entire Holocene. This decline appears to be continuing. In each of the first three times this pattern occurred, the declines in drift ice were associated with a period of major drought, and major vegetation change, in the Great Basin.

If the same pattern of major drought following a major high in ice rafting is repeated with the current decline, we may be on the descent side of the next drop into the next period of overall greater solar output and globally warmer climates. In the past, these warmer periods indicated by reduced ice rafting have lasted an average of three hundred to four hundred years. Following the Neoglacial peak, the drought in the Great Basin lasted six hundred to seven hundred years (Miller et al. 2001; chapter 3). If the past patterns continue, the Great Basin may experience approximately three hundred years of potentially drier climates. What this implies, is that it could be several centuries before the higher levels of moisture that were typical of the Neoglacial or Little Ice Age will return to the Great Basin.

Evident in the paleorecord has been a repeating pattern of large, abrupt, and widespread climate changes involving threshold shifts (Alley et al. 2003). These occur because climatic perturbations are amplified by feedback systems that can include their own sources of persistence, plus an ability to spread the anomalies across large regions. The faster the changes, the greater the likelihood of harmful impacts (Alley et al. 2003). Any factors causing additional forcing of a climate system also increases the probability of a threshold crossing. As a result, human influences on global warming could be increasing the probability of triggering such a large, abrupt event with potential serious ecological and economic impacts. The more rapid the global warming, the greater the likelihood of an abrupt change, and the greater the unpredictability of the possible changes (Alley et al. 2003). Such abrupt changes would involve the interactions of natural and anthropogenic causes.

A major contributing factor in future changes is the possible climatic effects of the anthropogenically driven increases in CO_2 and other greenhouse gasses in the atmosphere. As CO_2 levels rise, plant responses to changing climate may vary in ways that have not occurred in the past (Van de Water et al. 1994; Smith et al. 2000). The precise effects of increasing CO_2, particularly its potential contribution to a warmer climate, are still largely unquantified. It is doubtful, however, that the outcome will be to shorten the length of the warmer, and possibly drier period in the Great Basin we appear to be entering, or limit the duration of this trend. Models that incorporate the CO_2 increases indicate the possibility of an extended interglacial with levels of warming equaling the warmest phases of the Pleistocene (Berger and Loutre 2002). The possibility of a threshold shift in climate resulting from these forcings needs better study (Alley et al. 2003).

Ecological impacts from the recent changes in climate are also becoming evident (Walther et al. 2002). A contributing factor in the projection of future changes is the ongoing, worldwide interchange of species. This interchange may represent the most significant factor affecting Great Basin ecosystems because these species have the potential to permanently alter community dynamics (D'Antonio and Vitousek 1992). Attempts are now being made to project what the effects of these interactions will be on future distributions of major species and communities (Benson et al. 1998; Dale et al. 2000; Smith et al. 2000; Bachelet et al. 2001; Hansen et al. 2001).

The use of vegetative community reconstructions or models developed to represent Little Ice Age vegetation to determine restoration goals will increasingly become a heuristic exercise. As we proceed through the twenty-first century, it will be increasingly necessary to determine both the restoration procedures to be used and the goals for that restoration based on the environmental conditions of the present and those expected in the future. Attempts to restore Great Basin communities to what existed in the past (i.e., the Little Ice Age), represent an inaccurate characterization of what is possible now and in the future (Kloor 2000), and will seldom be successful (Tausch 1996; Millar 1997; Millar and Woolfenden 1999).

Acknowledgments

This work was supported in part by the funding of the Ecosystem Management Project, USDA Forest Service, Rocky Mountain Research Station. We would also like to acknowledge the very helpful discussions with Connie Millar of the Pacific Southwest Research Station during the development of this manuscript, and the thoughtful reviews of S. Kitchen, R. Nowak, J. Tainter, and R. Periman, which substantially improved the manuscript.

LITERATURE CITED

Agenbroad, L. D. 1978. Buffalo jump complexes in Owyhee County, Idaho. *Plains Anthropologist* 23:313–321.

Alley, R. B., J. Marotzke, W. D. Nordhaus, J. T. Overpeck, D. M. Peteet, R. A. Pielke Jr., R. T. Pierrehumbert, P. B. Rhines, T. F. Stocker, L. D. Talley, and J. M. Wallace. 2003. Abrupt climate change. *Science* 299:2005–2010.

Anderson, R. S., S. J. Smith, and P. A. Koehler. 1997. Distribution of sites and radiocarbon dates in the Sierra Nevada: Implications for paleoecological prospecting. *Radiocarbon* 39:121–137.

Bachelet, D., R. P. Neilson, J. M. Lenchan, and R. J. Drapek. 2001. Climate change effects on vegetation distribution and carbon budgets in the United States. *Ecosystems* 4:164–185.

Bailey, R. G., P. E. Avers, T. King, W. H. McNab, eds. 1994. *Ecoregions and subregions of the United States.* Map (scale 1:7,500,000). USDA Forest Service, Washington D.C.

Benson, L., M. Kashgarian, R. Rye, S. Lund, F. Paillet, J. Smoot, C. Kester, S. Mensing, D. Meko, and S. Lindström. 2002. Holocene multidecadal and multicentennial droughts affecting northern California and Nevada. *Quaternary Science Reviews* 21:659–682.

Benson, L. B., J. Burdett, M. Kashgarian, S. Lund, and S. Mensing. 1997. Nearly synchronous climate change in the Northern Hemisphere during the last glacial termination. *Nature* 388:263–265.

Benson, L. V., M. Kashgarian, and M. Rubin. 1995. Carbonate deposition, Pyramid Lake subbasin, Nevada, 2: Lake levels and polar jet stream positions reconstructed from radiocarbon ages and elevations of carbonates (Tufas) deposited in the Lahontan Basin. *Palaeogeography, Palaeoclimatology, Palaeoecology* 117:1–30.

Benson, R. S., S. W. Hostetler, P. J. Bartlein, and K. H. Anderson. 1998. *A strategy for assessing potential future changes in climate, hydrology, and vegetation in the western United States.* Circular 1153, U.S. Department of the Interior, U.S. Geological Survey, Washington, D.C.

Berger, A., and M. F. Loutre. 2002. An exceptionally long interglacial ahead? *Science* 297:1287–1288.

Betancourt, J. L. 1996. Long and short-term climate influences on southwestern shrublands. Pp. 5–9 in Proceedings: *Shrubland ecosystem dynamics in a changing environment,* compiled by J. R. Barrow, E. D. McArthur, R. E. Sosebee, R. J. Tausch; May 23–25, 1995. Las Cruces, N.M. General Technical Report INT-338, USDA Forest Service, Intermountain Research Station. Ogden, Utah.

Betancourt, J. L., E. A. Pierson, K. A. Rylander, J. A. Fairchild-Parks, and J. S. Dean. 1993. Influence of history and climate on New Mexico piñon-juniper woodlands. Pp. 42–62 in Proceedings: *Managing pinyon-juniper ecosystems for sustainability and social needs*, compiled by E. F. Aldon and D. W. Shaw. General Technical Report RM-236, USDA Forest Service, Rocky Mountain Forest and Range Experiment Station. Fort Collins, Colo.

Betancourt, J. L., T. VanDevender, and P. S. Martin. 1990. *Packrat middens: The last 40,000 years of biotic change.* Tucson: University of Arizona Press.

Billings, W. G. 1951. Vegetational zonation in the Great Basin of western North America. in *Les Bases ecologiques de la regeneration de la vegetation des zones arides.* International Union de la Societe biologique, Series B, 9:101–122.

Bond, G., B. Kromer, J. Beer, R. Muscheler, M. N. Evans, W. Showers, S. Hoffmann, R. Lotti-Bond, I. Hajdas, and G. Bonani. 2001. Persistent solar influence on North Atlantic climate during the Holocene. *Science* 294:2130–2136.

Bond, G., W. Showers, M. Cheseby, R. Lotti, P. Almasi, P. deMinocal, P. Priore, H. Cullen, I. Hajdas, G. Bonani. 1997. Pervasive millennial-scale cycle in North Atlantic Holocene and glacial climates. *Science* 278:1257–1266.

Born, S. M. 1972. *Late Quaternary history, deltaic sedimentation, and mudlump formation at Pyramid Lake, Nevada.* A report from the Desert Research Institute, University of Nevada, Reno.

Bradbury, J. P., R. M. Forestor, and R. S. Thompson. 1989. Late Quaternary paleolimnology of Walker Lake, Nevada. *Journal of Paleolimnology* 1:249–267.

Bradley, R. 2000. One thousand years of climate change. *Science* 288:1353–1355.

Briffa, K. A., and T. J. Osborn. 2002. Blowing hot and cold. *Science* 295:2227–2228.

Broecker, W. S. 2001. Was the Medieval Warm Period global? *Science* 291:1497–1499.

Brown, W. J., S. G. Wells, Y. Enzel, R. Y. Anderson, and L. D. McFadden. 1990. The late Quaternary history of pluvial Lake Mojave–Silver Lake and Soda Lake Basins, California. Pp. 55–72 in *At the end of the Mojave: Quaternary Studies in the eastern Mojave Desert.* Mojave Desert Quaternary Research Center Symposium, May 18–21, 1990. Special Publication of the San Bernardino County Museum Association, San Bernardino, Calif.

Bunting, S.C. 1994. Effects of fire on juniper woodland ecosystems. Pp. 53–55 in Proceedings: *Ecology and management of annual rangelands*, compiled by S. B. Monsen and S. G. Kitchen. General Technical Report INT-313, USDA Forest Service, Intermountain Research Station. Ogden, Utah.

Butler, R. R. 1978. Bison hunting in the desert West before 1800: The paleoecological potential and the archaeological reality. *Plains Anthropologist* 23:106–112.

Chambers, J. C., K. Farleigh, R. J. Tausch, J. R. Miller, D. Germanoski, D. Martin, and C. Nowak. 1998. Understanding long- and short-term changes in vegetation and geomorphic processes: The key to riparian restoration? Pp. 101–110 in *Proceedings of AWRA Specialty Conference, Rangeland Management and Water Resources*, edited by D. F. Potts. TPS-98-1. Herndon, Va.: American Water Resources Association.

Chatters, J. C., and K. A. Hoover. 1992. Response of the Columbia River fluvial system to Holocene climate change. *Quaternary Research* 37:42–59.

Clark, D. H., and A. R. Gillespie. 1997. Timing and significance of the late-glacial and Holocene cirque glaciation in the Sierra Nevada, California. *Quaternary International* 38/39:21–38.

Cole, K. L. 1985. Past rates of change, species richness, and a model of vegetational inertia in the Grand Canyon, Arizona. *Quaternary Research* 25:392–400.

Cordell, L. S., and G. J. Gumerman, eds. 1989. *Cultural interaction in the prehistoric Southwest: Dynamics of Southwest prehistory.* Washington, D.C.: Smithsonian Institution Press.

Creque, J. A. 1996. An ecological history of Tintic Valley (Juab County), Utah. PhD. diss. Utah State University, Logan.

Dale, V. D., L. A. Joyce, S. McNulty, and R. P. Neilson. 2000. The interplay between climate change, forests, and disturbances. *Science of the Total Environment* 262:201–204.

D'Antonio, C. M., and P. M. Vitousek. 1992. Biological invasions by exotic grasses, the grass/fire cycle, and global change. *Annual Review of Ecology and Systematics* 23:63–87.

Davis, J. O. 1982. Bits and pieces: The last 35,000 years in the Lahontan area. Society *American Archeology* 2:53–75.

Davis, M. B. 1986. Climatic instability: Time lags, and community disequilibrium. Pp. 269–284 in *Community ecology,* edited by J. Diamond and T. J. Case. New York: Harper and Row.

Davis, O. K. 1994. The correlation of summer precipitation in the southwestern USA with isotopic records of solar activity during the Medieval Warm Period. *Climate Change* 26:271–287.

Dean, J. S. 1996. Kayenta Anasazi settlement transformations in northeastern Arizona: A.D. 1150 to 1350. Pp. 29–47 in *The prehistoric pueblo world, AD 1150–1350,* edited by M. A. Adler. Tucson: University of Arizona Press.

Denton, G. H., and W. Karlén. 1973. Holocene climatic variations—their pattern and possible cause. *Quaternary Research* 3:155–205.

Enzel, Y., W. J. Brown, R. Y. Anderson, L. D. McFadden, and S. G. Wells. 1992. Short-duration Holocene lakes in the Mojave River drainage basin, southern California. *Quaternary Research* 38:60–73.

Enzel, Y., D. R. Cayan, R. Y. Anderson, and S. G. Wells. 1989. Atmospheric circulation during Holocene lake stands in the Mojave Desert: Evidence of regional climate change. *Nature* 341:44–47.

Esper, J., E. R. Cook, and F. H. Schweingruber. 2002. Low-frequency signals in long tree-ring chronologies for reconstructing past temperature variability. *Science* 295:2250–2252.

Farquhar, G. D. 1997. Carbon dioxide and vegetation. *Science* 278:1411.

Furgurson, E. B., and G. F. Mobley. 1992. Lake Tahoe: Playing for high stakes. *National Geographic* 181:112–132.

Grayson, D. 1993. *The desert's past: A natural prehistory of the Great Basin.* Washington, D.C.: Smithsonian Institution Press.

Grove, J. M., and R. Switsur. 1994. Glacial geological evidence for the Medieval Warm Period. *Climate Change* 26:143–169.

Gruell, G. E. 1999. Historical and modern roles of fire in pinyon-juniper. Pp. 24–28 in *Proceedings: Ecology and management of pinyon-juniper communities within the Interior West,* compiled by S. B. Monsen and R. Stevens. Proceedings RMRS P-9. USDA Forest Service, Rocky Mountain Research Station. Ogden, Utah.

Hall, S. A. 1990. Channel trenching and climatic change in the southern U.S. Great Plains. *Geology* 18:342–345.

Hansen, A. J., R. P. Neilson, V. H. Dale, C. H. Flather, C. R. Iverson, D. J. Currie, S. Shafer, R. Cook, and P. J. Bartlein. 2001. Global change in forests: Responses of species, communities, and biomass. *BioScience* 51:765–779.

Hiller, A., T. Boettger, and C. Kremenetski. 2001. Mediaeval climatic warming recorded by radiocarbon dated alpine tree-line shift on the Kola Peninsula, Russia. *The Holocene* 11:491–497.

Holmes, R. L., R. K. Adams, and H. C. Fritts. 1986. *Tree-ring chronologies of western North America: California, eastern Oregon and northern Great Basin and procedures used in the chronology development work including users manual for computer programs.* COFECHA and ARSTAN, Chronology Series 6, Tree-Ring Laboratory. Tucson, Ariz.

Jackson, S. T., M. E. Lyford, and J. L. Betancourt. 2002. Four thousand year record of woodland vegetation from Wind River Canyon, central Wyoming. *Western North American Naturalist* 62:405–413.

Jennings, S. A., and D. L. Elliot-Fisk. 1993. Packrat midden evidence of late Quaternary vegetation change in the White Mountains, California-Nevada. *Quaternary Science* 39:214–221.

Kaiser, J. 2001. The other global pollutant: Nitrogen proves tough to curb. *Science* 294:1268–1269.

Kerr, R. A. 2001. A variable sun paces millennial climate. *Science* 294:1431–1433.

Kinney, W. C. 1996. Conditions of rangelands before 1905. Pp. 31–45 in *Sierra Nevada Ecosystem Project: Final report to Congress*, Vol. 2, chap. 3. University of California, Davis, Centers for Water and Wildland Resources.

Kloor, K. 2000. Returning America's forests to their natural roots. *Science* 287:573–575.

Konrad, S. K., and D. H. Clark. 1998. Evidence of an early Neoglacial glacier advance from rock glaciers and lake sediments in the Sierra Nevada, California, USA. *Arctic and Alpine Research* 30:272–284.

LaMarche, V. C. 1973. Holocene climatic variations inferred from treeline fluctuations in the White Mountains, California. *Quaternary Research* 3:632–660.

———. 1974. Paleoclimatic inferences from long tree-ring records. *Science* 183:1043–1048.

Leavitt, S. W. 1994. Major wet interval in White Mountains Medieval Warm Period evidenced in $\delta^{13}C$ of bristlecone pine rings. *Climate Change* 26:299–307.

Lewin, R. 1985. Plant communities resist climatic change. *Science* 228:165–166.

Lindstrom, S. 1990. Submerged tree stumps as indicators of mid-Holocene aridity in the Lake Tahoe Basin. *Journal of California and Great Basin Anthropology* 12:146–157.

Lloyd, A. H., and L. J. Graumlich. 1997. Holocene dynamics of treeline forests in the Sierra Nevada. *Ecology* 78:1199–1210.

Luckman, B. H. 1990. Mountain areas and global change: A view from the Canadian Rockies. *Mountain Research and Development* 10:183–195.

Lyford, M. E., J. L. Betancourt, and S. T. Jackson. 2002. Holocene vegetation and climate history of the northern Bighorn Basin, southern Montana. *Quaternary Research* 58:171–181.

Madsen, D. B. 1985. Two Holocene pollen records for the central Great Basin. Pp. 113–126 in *Late Quaternary vegetation and climates of the American Southwest*, edited by B. F. Jacobs, P. L. Fall, and O. K. Davis. Contribution 16. American Association of Stratigraphic Palynologists.

Madsen, D. B., D. Rhode, D. K. Grayson, J. M. Broughton, S. D. Livingston, J. Hunt, J. Quade, D. N. Schmitt, and M. W. Shaver III. 2001. Late Quaternary environmental change in the Bonneville Basin, western USA. *Palaeogeography, Palaeoclimatology, Palaeoecology* 167:243–271.

Mehringer, P. J., Jr. 1977. Great Basin late Quaternary environments and chronology. Pp. 113–167 in *Models and Great Basin prehistory: A symposium*, edited by D. Fowler. Desert Research Institute Publications in Social Sciences No. 12, Desert Research Institute. Reno, NV.

———. 1985. Late-quaternary pollen records from the interior Pacific Northwest and northern Great Basin of the United States. Pp. 167–189 in *Pollen records of late-Quaternary North American sediments*, edited by V. M. Bryant. Jr. and R. G. Holloway. Dallas, Tex.: American Association of Stratigraphic Palynologists.

———. 1987. *Late Holocene environments on the northern periphery of the Great Basin.* Final Report. U.S. Department of the Interior, Bureau of Land Management, Portland, Ore.

Mehringer, P. J., Jr., and C. N. Warren. 1976. Marsh, dune, and archeological chronologies, Ash Meadows, Amargosa Desert, Nevada. *Research Report of the Nevada Archeological Survey* 6:120–150.

Mehringer, P. J., Jr., and P. E. Wigand. 1990. Comparison of late Holocene environments from woodrat middens and pollen. Pp. 294–325 in *Packrat middens: The last 40,000 years of biotic change*, edited by J. L. Betancourt, T. R. VanDevender, and P. S. Martin. Tucson: University of Arizona Press.

Meko, D. M., M. D. Therrell, C. H. Baisan, and M. K. Hughes. 2001. Sacramento River flow reconstructed to AD 869 from tree rings. *Journal of the American Water Resources Association* 37:1029–1040.

Mensing, S. A. 2001. Late glacial and early Holocene vegetation and climate change near Owens Lake, eastern California. *Quaternary Research* 55:55–67.

Millar, C. I. 1996. Tertiary vegetation history. Pp. 71–109 in Sierra Nevada Ecosystem Project: Final Report to Congress, Vol. 2, chap. 5. University of California, Davis, Centers for Water and Wildland Resources.

———. 1997. Comments on historical variation and desired condition as tools for terrestrial landscape analysis. Pp. 105–131 in *What is watershed stability?*, edited by S. Sommarstrom. Proceedings of the sixth Biennial Watershed Management Conference. Report No. 92. Oct. 23–25, 1996; Lake Tahoe, California/Nevada. University of California, Davis, Water Resources Center.

Millar, C. I., and W. B. Woolfenden. 1999. The role of climate change in interpreting historical variability. *Ecological Applications* 9:1207–1216.

Miller, J., D. Germanoski, K. Waltman, R. Tausch, and J. Chambers. 2001. Influence of late Holocene hillslope processes and landforms on modern channel dynamics in upland watersheds of central Nevada. *Geomorphology* 38:373–391.

Miller, R. E., and R. J. Tausch. 2001. The role of fire in pinyon and juniper woodlands: A descriptive analysis. Pp. 15–30 in *Proceedings of the Invasive Species workshop: The role of fire in the control and spread of invasive species*, edited by K. E. M. Galley and T. P. Wilson. Miscellaneous Publication No. 11, Fire Conference 2000: The first national congress on fire ecology, prevention, and management. Tall Timbers Research Station, San Diego, CA, Nov. 27–Dec. 1, 2000. Tallahassee, Fla.

Miller, R. E., and P. E. Wigand. 1994. Holocene changes in semiarid pinyon-juniper woodlands. *BioScience* 44:465–474.

Naftz, D. L., R. W. Klusman, R. L. Michel, P. F. Shuster, M. M. Reddy, H. E. Taylor, T. M. Yanosky, and E. A. McConnaughey. 1996. Little Ice Age evidence from a south-central North American ice core. *Arctic and Alpine Research* 28:35–41.

Nowak, C. L., R. S. Nowak, R. J. Tausch, and P. E. Wigand. 1994a. Tree and shrub dynamics in northwestern Great Basin woodland and shrub steppe during the late-Pleistocene and Holocene. *American Journal of Botany* 81:265–277.

———. 1994b. A 30,000 year record of vegetation dynamics at a semi-arid locale in the Great Basin. *Journal of Vegetation Science* 5:579–590.

Oviatt, C. G. 1997. Lake Bonneville fluctuations and global climate change. *Geology* 25:155–158.

Pyke, C. B. 1972. *Some meteorological aspects of the seasonal distribution of precipitation in the western United States and Baja California*. Contribution 139. University of California, Los Angeles, Water Resources Center.

Rhode, D. 2000. Packrat middens as a tool for reconstructing historic ecosystems. Pp. 257–293 in *The historical ecology handbook: A restorationist's guide to reference ecosystems*. Washington, D.C.: Island Press.

Schroedl, G. F. 1973. *The archeological occurrence of bison in the southern Plateau*. Report of Investigation 51. Washington State University, Lobo, Laboratory of Anthropology. Pullman, Wash.

Scuderi, L. A. 1987. Late Holocene upper timberline variation in the southern Sierra Nevada. *Nature* 325:242–244.

Smith, S. D., T. E. Huxman, S. F. Zitzer, T. N. Charlet, D. C. Housman, J. S. Coleman, L. K. Fenstermaker, J. R. Seeman, and R. S. Nowak. 2000. Elevated CO_2 increases productivity and invasive species success in an arid system. *Nature* 408:79–82.

Spaulding, W. G. 1985. *Vegetation and climates of the last 45,000 years in the vicinity of the Nevada Test Site, south-central Nevada*. United States Geological Survey Professional Paper 1329.

———. 1990. Vegetational and climatic development of the Mojave Desert: The last glacial maximum to present. Pp. 166–199 in *Packrat Middens: The Last 40,000 Years of Biotic Change*, edited by J. L. Betancourt, T. R. VanDevender, and P. S. Martin. Tucson: University of Arizona Press.

———. 1991. A middle Holocene vegetation record from the Mojave Desert of North America and its paleoclimatic significance. *Quaternary Research* 35:427–437.

Stine, S. 1990a. Late Holocene fluctuations of Mono Lake, eastern California. *Palaeogeography, Palaeoclimatology, Palaeoecology* 78:333–381.

———. 1990b. Past climate at Mono Lake. *Nature* 345:391.

———. 1994. Extreme and persistent drought in California and Patagonia during medieval time. *Nature* 369:546–549.

———. 1996. Climate 1650–1850. Pp. 25–30 in *Sierra Nevada ecosystem project: Final report to congress*, Vol. 2, chap. 2. University of California, Davis, Centers for Water and Wildland Resources.

Stutz, H. C. 1984. *Atriplex* hybridization in western North America. Pp. 25–27 *in Proceedings, symposium on the biology of Atriplex and related chenopods*, compiled by A. R. Tiedemann, E. D. McArthur, H. C. Stutz, R. Stevens, and K. L. Johnson. General Technical Report INT-172, USDA Forest Service, Intermountain Forest and Range Experiment Station. Ogden, Utah.

Swetnam, T. W. 1993. Fire history and climate change in giant sequoia groves. *Science* 262:885–889.

Tausch, R. J. 1996. Past changes, present and future impacts, and the assessment of community or ecosystem condition. Pp. 97–101 in Proceedings: *Shrubland ecosystems dynamics in a changing environment*, compiled by J. R. Barrow, E. D. McArthur, R. E. Sosebee, and R. J. Tausch. General Technical Report INT-338, USDA Forest Service, Intermountain Research Station. Ogden, Utah.

———. 1999. Historic pinyon and juniper woodland development. Pp. 12–19 in Proceedings: *Ecology and management of pinyon-juniper communities within the Interior West*, compiled by S. B. Monsen and R. Stevens. Proceedings RMRS P-9, USDA Forest Service, Rocky Mountain Research Station. Ogden, Utah.

Tausch, R. J., P. E. Wigand, and J. W. Burkhardt. 1993. Viewpoint. Plant community thresholds, multiple steady states, and multiple successional pathways: Legacy of the Quaternary? *Journal of Range Management* 46:439–447.

Tausch, R. J., and C. L. Nowak. 2000. Influences of Holocene climate and vegetation change on present and future community dynamics. In Desert Technology, special issue, *Journal of Arid Land Studies* 10S:5–8.

Thompson, R. S. 1984. Late Pleistocene and Holocene environments in the Great Basin. PhD. diss. University of Arizona, Tucson.

———. 1990. Late Quaternary vegetation and climate in the Great Basin. Pp. 200–239 in *Packrat middens: The last 40,000 years of biotic change*, edited by J. L. Betancourt, T. VanDevender, and P. S. Martin. Tucson: University of Arizona Press.

Thompson, R. S., and R. R. Kautz. 1983. Paleobotany of Gatecliff Shelter: Pollen analysis. Pp. 136–157 in *The archaeology of Monitor Valley. 2. Gatecliff Shelter*, edited by D. H. Thomas. Anthropological Papers of the American Museum of Natural History. Vol. 59.

Thompson, R. S., C. Whitlock, P. J. Bartlein, S. P. Harrison, and W. G. Spaulding. 1993. Climatic changes in the western United States since 18,000 yr BP. Pp. 468–513 in *Global climates since the last glacial maximum*, edited by H. E. Wright Jr., J. E. Kutzbach, T. Webb III, W. F. Ruddiman, F. A. Street-Perrott, and P. J. Bartlein. Minneapolis: University of Minnesota Press.

VanDevender, T. R., and W. G. Spaulding. 1979. Development and climate in the southwestern United States. *Science* 204:701–710.

Van de Water, P. K., S. W. Leavitt, and J. L. Betancourt. 1994. Trends in stomatal density and $^{13}C/^{12}C$ ratios of *Pinus flexilis* needles during the last glacial-interglacial cycle. *Science* 264:239–243.

Walker, M. J. C. 2001. Rapid climate change during the last glacial-interglacial transition: Implications for stratigraphic subdivision, correlation and dating. *Global and Planetary Change* 30:59–72.

Walther, G., E. Post, P. Convey, A. Menzal, C. Parmesan, T. J. C. Beebee, J. Fromentin, D. Hoegh-Guldberg, and F. Bairlen. 2002. Ecological responses to recent climate change. *Nature* 416:389–395.

Wells, P. K. 1983. Paleobiography of montane islands in the Great Basin since the last glaciopluvial. *Ecological Monographs* 53:341–382.

West, N. E., R. J. Tausch, K. H. Rea, and P. T. Tueller. 1978. Taxonomic determination, distribution, and ecological indicator values of sagebrush within the pinyon-juniper woodlands of the Great Basin. *Journal of Range Management* 31:87–92.

West, N. E., R. J. Tausch, and P. T. Tueller. 1998. *A management-oriented classification of pinyon-juniper woodlands of the Great Basin*. General Technical Report RMRS-12, USDA Forest Service, Rocky Mountain Research Station. Ogden, Utah.

Whitlock, C., and P. J. Bartlein. 1997. Vegetation and climate change in northwest America during the past 125k yr. *Nature* 388:57–61.

Wigand, P. E. 1987. Diamond Pond, Harney County, Oregon: Vegetation history and water table in the eastern Oregon Desert. *Great Basin Naturalist* 47:427–458.

Wigand, P. E., M. L. Hemphill, S. Sharpe, and S. Patra. 1995. Great Basin semi-arid woodland dynamics during the late Quaternary. Pp. 51–70 in *Proceedings: Climate change in the four corners and adjacent regions: Implications for environmental restoration and land-use planning*, edited by W. J. Waugh; Sept. 12–14, 1994; Mesa State College, Grand Junction, Colo.: U.S. Department of Energy.

Wigand, P. E., and P. J. Mehringer Jr. 1985. Pollen and seed analyses. In The archaeology of Hidden Cave, Nevada, edited by D. H. Thomas, *Anthropological Papers of the American Museum of Natural History* 61:108–124.

Wigand, P. E., and C. L. Nowak. 1992. Dynamics of northwest Nevada communities during the last 30,000 years. In *The history of water: eastern Sierra Nevada, Owens Valley, White-Inyo Mountains*, edited by C. A. Hall, V. Doyle-Jones, and B. Widawski. White Mountain Research Station Symposium. Vol. 4.

Wigand, P. E., and D. Rhode. 2002. Great Basin vegetation history and aquatic systems: The last 150,000 years. Pp. 309–367 in *Great Basin aquatic ecosystems history*, edited by R. Hershler, D. B. Madsen, and D. R. Currey. Smithsonian Contributions to Earth Sciences 33. Washington, D.C.: Smithsonian Institution Press.

Wigand, P. E., and M. R. Rose. 1990. *Calibration of high frequency pollen sequences and tree-ring records*. Proceedings of the international high level radioactive waste management conference and exposition, April 1990, Las Vegas, Nev.

Winward, A. H. 1969. Taxonomic and ecological relationships of the big sagebrush complex in Idaho. Ph.D. diss. University of Idaho, Moscow. University Microfilms, Ann Arbor, Mich.

Woolfenden, W. B. 1996. Quaternary vegetation history. Pp. 47–70 in *Sierra Nevada ecosystem project: Final report to congress*, Vol. 2, chapter 4. University of California, Davis, Centers for Water and Wildland Resources.

Zic, M., R. M. Negrini, and P. E. Wigand. 2002. Evidence of synchronous climate change across the Northern Hemisphere between the North Atlantic and northwestern Great Basin. *Geology* 30:635–638.

Chapter 3

Fluvial Geomorphic Responses to Holocene Climate Change

JERRY R. MILLER, KYLE HOUSE, DRU GERMANOSKI,
ROBIN J. TAUSCH, AND JEANNE C. CHAMBERS

Geomorphologists widely accept the premise that climatic disturbances periodically altered river processes and form in the western United States prior to Anglo-American settlement. It is, therefore, surprising that the ability of climatic perturbations to produce significant fluvial geomorphic responses during the past 150 years has been largely ignored. The general perception by many ecologists, land managers, and environmentalists, as well as geomorphologists, is that recent changes in channel morphology, particularly incision associated with adjustments in gradient, are the direct result of human activities (Cooke and Reeves 1976; McFadden and McAuliffe 1997). In fact, many believe that human activities represent the universal cause of channel change in the western United States today (see, for example, Jacobs 1991). This perception is based on several factors. First, numerous studies have shown that causal or mechanistic linkages exist between human activities and channel degradation (Gifford and Hawkins 1978; Trimble and Mendel 1995), leaving little doubt that anthropogenic disturbances can dramatically impact fluvial process and form. Second, changes in climate are perceived to be too insignificant in comparison to the magnitude of human disturbances to produce changes in channel characteristics. From this perspective, climatic disturbances are overwhelmed by the more numerous and more disruptive alterations to the landscape by humans. Finally, many observers fail to recognize the dynamic nature of rivers prior to the historical period (Miller and Kochel 1999).

While there is no question that human activities have led to significant changes in channel morphology and process rates, a complete understanding of the factors controlling fluvial systems in any given region requires an assessment of the impacts of both anthropogenic disturbances

49

and climatic change on fluvial processes. This chapter examines the relations between minor changes in climate and fluvial geomorphic processes during the Holocene within drainage systems of vastly differing spatial scales of the central Great Basin. The analysis provides insights into the nature of future geomorphic adjustments to both natural and anthropogenic disturbances. From this perspective, the uniformitarian principle (the present is the key to the past) is reversed so that the past becomes the key to the present and future (Knighton 1998). In addition, the influence of past geomorphic processes, and the landforms that were produced by them, on modern channel dynamics is examined. Ultimately, the conclusions presented here will lead to a more complete understanding of the processes, and process controls, on fluvial systems in the central Great Basin, an understanding that is needed for the sound development of restoration and management strategies.

Previous Investigations of Geomorphic Responses to Climate Change

The current paradigm of fluvial mechanics asserts that gradients, cross-sectional dimensions, patterns (straight, meandering, braided), and planimetric configurations (e.g., sinuosity) of alluvial channels are adjusted over a period of years to most efficiently transport the water and sediment delivered to the reach from upstream areas (Ritter et al. 2002). Thus, the hydrologic and sedimentologic regimes are the primary controls on channel morphology. Discharge and sediment load are not independent variables but are ultimately governed by climate, vegetation, geology, and physiography of the watershed. Over short time scales (10^1 to 10^2 years), geology and physiography are unlikely to be significantly altered (Schumm and Lichty 1965). Changes in climate, however, can affect channel morphology by directly altering discharge and sediment-load characteristics or by indirectly altering upland vegetation that in turn influences runoff and sediment yields.

Investigations concerning the effects of climatic change on process and form date back to the turn of the century (e.g., Dutton 1882). These climatic-geomorphic studies have generated a number of conceptual models describing the nature of river adjustments to alterations in climate. The majority of these investigations are based on temporal linkages between alterations in channel morphology and past changes in temperature and precipitation. Documentation of channel metamorphosis is assessed using a combination of historic measurements, morphologic data (e.g., terraces and abandoned channels), stratigraphic information that records the timing and extent of erosional and depositional events, and sedimentologic data that can be used to assess changes in flow regime. Almost without exception, these data sets are complex and fragmented, yielding spatially and temporally incomplete records of channel change (Schumm and Brakenridge 1987). Depending on the time frame under consideration, changes in climate are documented using historical data or a variety of climatic proxy indicators, including pollen, packrat middens, tree rings, and stable isotopes as described in chapter 2. The climatic record is also temporally and spatially incomplete. Thus, elucidating the impacts of climate change on fluvial systems has proven difficult. Nonetheless, the body of knowledge on the responses of fluvial systems to climate change has increased dramatically during the past two decades for a wide variety of climatic, geologic, and physiographic settings. The emerging perception is that geomorphic responses to climatic disturbances are highly complex at the regional scale (Bull 1991; Knighton 1998). Some rivers may absorb minor shifts in climate and remain in a state of equilibrium, while others go through a complete metamorphosis in process and form. These contrasting responses are at least partly attributed to variations in landscape (or landform) sensitivity (Gerrard 1993).

Sensitivity has been defined in a number of different ways, depending on the objectives of the investigation (see, for example, Downs and Gregory 1993). In general, however, it describes the tendency of landforms to respond to an environmental disturbance and attain a new equilibrium state (Schumm and Brakenridge 1987). It is related to the interaction of driving and resisting forces that regulate erosional and depositional processes and, thus, is controlled by such factors as the erosional resistance of the underlying bedrock and the channel forming materials, basin relief, and morphometry (e.g., drainage density, stream frequency, and Shreve magnitude), and watershed hydrology (Gerrard 1993).

Recent studies have also demonstrated that when geomorphic responses to climatic perturbations do occur, those responses are often inconsistent from one watershed to another (Schumm and Brakenridge 1987). For example, channels in one basin may be characterized by channel aggradation, whereas another may experience channel incision in response to similar changes in climate. These inconsistencies may be related to fundamental differences in the production and transport of sediment within the watersheds. For instance, on the basis of empirical data, Langbein and Schumm (1958) argued that basin sediment yields do not directly vary with effective precipitation because the effects of precipitation are moderated by the density, composition, and distribution of the local vegetation. Although actual values vary with the data set used, maximum sediment yields are obtained in the United States with an effective precipitation between 200 and 500 millimeters (fig. 3.1); sediment yields are reduced with either increases or decreases in effective precipitation from this peak as a function of vegetative cover (Schumm and Brakenridge 1987; Langbein and Schumm 1958). The Langbein-Schumm (1958) curve was never intended to be a predictive model of the potential impacts of climatic disturbances on rivers. Nonetheless, it suggests that a decrease in effective precipitation may either reduce or enhance sediment yields depending on the antecedent moisture conditions of the basin. Contrasting impacts of climate change on sediment generation and delivery may partially explain the inconsistent behavior of rivers to a single, unidirectional change in climate. Other causes for divergent responses are likely to be differences in geology, physiography, and basin evolutionary history.

Asynchronous episodes of erosion or deposition along the channel network have also been observed in a number of fluvial systems (Schumm and Brakenridge 1987). These differences occur because thresholds are crossed at different times and under different magnitudes of climatic and hydrologic change within different parts of the watershed. Perhaps the most significant variations have been observed between headwater and downstream reaches (Knighton 1998). Headwater areas, with their closer proximity to the sources of water and sediment production, their lower storage capacities, and their greater potential energy, tend to be more responsive than downstream reaches (see Balling and Wells 1990, for an example).

Although the responses of river systems to changes in climate may be highly varied and complex in nature, an increasing number of papers have argued that changes in climate can have a significant, if not an overriding, effect on modern channel dynamics (Bryan 1928; Karlstrom et al. 1974; Karlstrom 1983; Karlstrom and Karlstrom 1986; Knox 1984; Hereford 1986; Balling and Wells 1990; Chatters and Hoover 1992; Macklin and Lewin 1993; Rumsby and Macklin 1994; Kochel et al. 1997; McFadden and McAuliffe 1997; Fuller et al. 1998). As described earlier, many of these arguments are based on dated alluvial stratigraphic sequences that show a temporal correlation between depositional events and changes in climate as well as atmospheric circulation patterns. Macklin and Lewin (1993) present one of the most complete analyses of these stratigraphic as-

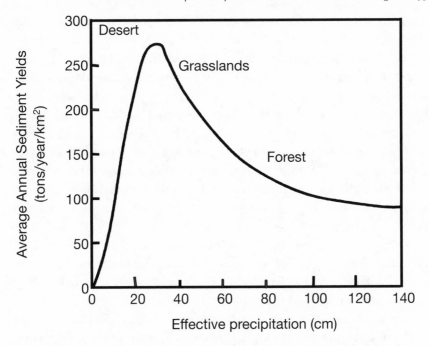

FIGURE 3.1. Variations in average annual sediment yield as a function of effective precipitation. W. B. Langbein and S. A. Schumm, "Yield of sediment in relation to mean annual precipitation," *Transactions of the American Geophysical Union* 36:1076–1084. Copyright 1958 American Geophysical Union. Reproduced/modified by permission of American Geophysical Union.

semblages. Using information published in fifty-nine papers on alluvial stratigraphy from Britain, they argue that there has been widespread, synchronous episodes of aggradation in the United Kingdom during that past five thousand years that correspond with climatic disturbances. They go on to state that

> Although forest clearance and agricultural practices were undoubt-edly important for initiating erosion, significant redistribution of this material and widespread alluviation generally occurred only during relatively short periods of abrupt climatic shift that were characterized by major changes in flood frequency and magnitude. Thus the British Holocene fluvial record can be considered to be an alluvial sequence that has been climatically driven but culturally blurred.

Analyses paralleling those of Macklin and Lewin (1993) in the United States have produced similar results (see, for example, Balling and Wells 1990; Miller and Kochel 1999).

Using a slightly different approach, McFadden and McAuliffe (1997) examined the timing and nature of channel responses in highly erodible terrain of the southern Colorado Plateau in the southwestern United States. The sites selected for study lacked evidence of grazing on hillslopes, and the geographical patterns of plants in the area were not those that are typically associated with livestock grazing, suggesting that livestock could not have caused the observed erosional and depositional events. These observations, in conjunction with other geomorphic and botanical data, led them to conclude that in areas underlain by highly sensitive geologic units, "even relatively minor climatic changes during the last one to two centuries have largely overwhelmed the effects of grazing in virtually every part of the landscape."

Differentiation of the effects of anthropogenic disturbance from climate change are complicated by the spatial scale at which investigations are conducted (Macklin and Lewin 1993). By necessity, instrumented experimental catchments, used to decipher fluvial geomorphic responses to disturbance, are small to ensure that accurate results are obtained. These analyses have generated conclusive data that human activities alter basin hydrology and sediment yields which ultimately lead to the crossing of geomorphic thresholds (Arnell 1989). However, extrapolating data from small watersheds to large systems is problematic. Thus, it is not clear that the dramatic hydrologic and geomorphic effects of human activities observed in smaller, instrumented catchments occur in larger systems (Newson and Lewin 1991) that exhibit greater water and sediment storage potentials. There is, in fact, some evidence to suggest that at larger spatial scales, minor climatic change has a relatively greater impact on geomorphic processes. Hereford (1984), for example, examined the discharge records and tree-ring width indices pertinent to the Little Colorado River. He found that there was an abrupt rise in precipitation that began in 1900 and that the channel at this time was characterized by frequent, large floods. This was followed by a semisystematic decline in precipitation and soil moisture that extended from 1910 to 1956. By the 1940s and early 1950s, annual discharge had decreased to approximately 57 percent of that of the earlier period and the frequency of large floods had declined sharply. The hydrologic and climatic patterns documented by Hereford (1984) led to changes in the dominant geomorphic processes operating within the Little Colorado River basin. Beginning at about 1905 and continuing until 1937, erosion dominated the system. Shortly after 1937, the erosional period ceased, channel width decreased, the floodplain aggraded, and vegetation began to stabilize the higher parts of the older channel. Floodplain stabilization and aggradation was related to a decline in precipitation, mean annual discharge, and the frequency of large floods. The significance of this study (and others, such as Balling and Wells 1990) is that hydrophysical links were made between climatic indices, hydrologic parameters, and geomorphic processes, suggesting that at relatively large spatial scales, climate change plays a significant role in controlling geomorphic processes in arid or semiarid environments.

The literature presented above illustrates that minor changes in climate can significantly impact modern channel dynamics and that it is unrealistic to assign either climate change or human activities as the universal cause for recent adjustments in channel morphology. Both are likely to be primary controls on channel processes, and the relative influence will undoubtedly vary with climatic regime, basin geology, physiography, vegetation, and land-use history. It follows, then, that the development of sound restoration and management strategies requires an assessment of fluvial processes within the catchment of interest.

Geomorphic Responses of Upland Watersheds to Late Holocene Disturbances

The mountain ranges in the central Great Basin are characterized by numerous, small (less than 100-square-kilometer) watersheds that terminate in coalescing alluvial fans that flank the intermountain basins. An analysis of landform morphology and the alluvial stratigraphic architecture of the valley fill within a number of these basins were used to develop a chronology of the erosional and depositional events that characterize the late Holocene. Moreover, the paleoclimatic data presented in the previous chapter provided the information necessary to assess the role of minor climatic change in altering both hillslope and channel processes during the past several thousand years.

Detailed stratigraphic and geomorphic analysis focused on seven upland watersheds including the Big Creek, Cottonwood Creek, Kingston Canyon, San Juan Creek, and Washington Creek basins in the Toiyabe Range, the Stoneberger Creek basin in the Toquima Range, and the Barley Creek basin within the Monitor Range (fig. 3.2). The elevation of these basins range from approximately 1,850 to 3,200 meters, with annual precipitation ranging from 20 centimeters at the basin mouth to 45 centimeters at upper elevations. At low to middle elevations, Wyoming big sage-

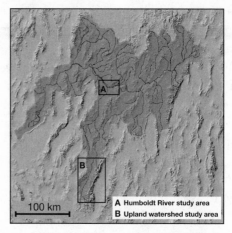

FIGURE 3.2. Location of the study areas described in the text. The Humboldt River watershed is indicated by darker shading. Also shown are the general locations of the upland watersheds and other sites mentioned in the text: PL = Pyramid Lake, RV = Ruby Valley, and LB = western margin of Lake Bonneville Basin.

brush (*Artemisia tridentata wyomingensis*) communities are interspersed with Utah juniper (*Juniperus osteosperma*) and single leaf pinyon (*Pinus monophylla*) woodlands. At higher elevations, mountain brush vegetation is dominant, including mountain big sagebrush (*A. tridentata vaseyana*) and limber pine (*P. flexilis*). The most prominent feature of the alluvial valley floors is the riparian corridor, characterized by relatively mesic vegetative communities that form the dynamic interface between the upland and aquatic systems. The riparian corridors are locally constricted by alluvial fan deposits constructed at the mouth of small tributaries that drain the hillslopes (fig. 3.3). Wetlands, or meadow complexes, preferentially occur upstream of these side-valley fan deposits (as they will be called in this chapter). Although these riparian areas comprise less than 2 percent of the total landscape, they contain a disproportionately large percentage of the region's biodiversity, providing habitat for both endemic and migrant species (Knopf et al. 1988).

The watersheds are underlain by a wide variety of rock types, the most prominent including volcanic lithologies (e.g., welded and nonwelded tuffs, rhyolites, and basalts), crystalline intrusives, and sedimentary-metasedimentary assemblages (Stewart and Carlson 1976; Kleinhampl and Ziony 1985). The tectonic environment is also highly variable as rates of uplift and tectonic activity vary both between ranges and along the mountain front of any given range (Wallace 1987; Dohrenwend et al. 1992).

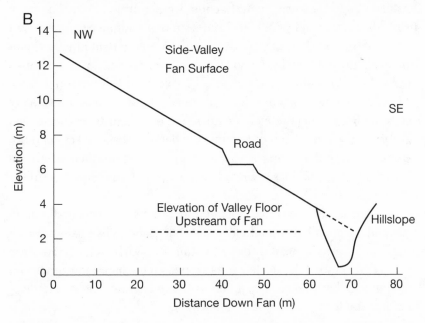

FIGURE 3.3. (A) Side-valley alluvial fan in Kingston Canyon, Nev., and (B) radial profile of the alluvial fan shown in A. Before the fan toe was truncated, fan deposits impinged on the opposing hillslope. (Figure modified from Miller et al. 2001.)

However, none of the late Holocene stratigraphic units within these basins are offset by late Quaternary fault scarps.

Methodology

The chronology of erosional and depositional events was determined by delineating, characterizing, and dating the alluvial stratigraphy of the mid- to late-Holocene valley fill exposed within 1- to 5-meter-high bank exposures along entrenched channels. The delineation of individual stratigraphic units was based on grain-size distribution, sediment color, clast lithology, degree of sediment weathering, and topographic-stratigraphic position. Once the units had been defined, the stratigraphic sections were mapped in detail to document the geometry of unit boundaries, unit and facies continuity, the location of collected radiocarbon and sediment samples, and the presence of surface or buried soils. The grain-size distribution of the greater-than-2-millimeter sediment fraction was determined using a modified Wolman (1954) approach, and the grain-size distribution of the less-than-2-millimeter sediment fraction was performed in the laboratory using wet-sieving and pipette techniques modified from Singer and Janitzky (1986). Soil profiles were described according to the methods and nomenclature put forth by the Soil Conservation Service (1981) and Birkeland (1984). Age control was provided by seventeen radiocarbon dates on organic sediments, peat, and charcoal collected from four different watersheds and five distinct stratigraphic units. The approximate timing of channel incision was determined by dating the deposits that bracket unconformities. Because the dating is based on materials contained within the alluvial deposits, episodes of aggradation are more accurately determined than periods of incision.

In order to quantify the topographic relations between stratigraphic units and document the morphology of selected landforms, cross-valley and longitudinal profiles were surveyed within the watersheds. In addition, stream reaches were delineated and mapped on the basis of channel form and gradient, bed and bank material size, number and height of terraces, and depth of incision below the valley floor.

Relations Between Geomorphic Processes and Late Holocene Climatic Variations

Geomorphic and stratigraphic data indicate that the alluvial architecture of the valley fill is similar within the upland watersheds that were examined in detail within the Toiyabe, Toquima, and Monitor Ranges. More-

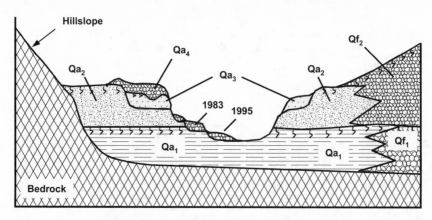

FIGURE 3.4. Schematic diagram of mid- to late-Holocene stratigraphy within the upland study basins. (Figure from Miller et al. 2001.)

over, extensive field reconnaissance suggests that while the magnitude of erosional and depositional events as defined by unit thickness, areal exposure, and distribution varies between basins, the stratigraphy observed within the study basins is similar in overall structure to that found in a large number of upland watersheds of the central Great Basin.

The mid- to late-Holocene valley fill consists of seven alluvial stratigraphic units that have been named, from oldest to youngest, Quaternary alluvial unit 1 through Quaternary alluvial unit 7 (Qa_1–Qa_7). In addition, the valley fill locally contains two alluvial fan units (Qf_1 and Qf_2) that represent materials that were delivered to the valley floor from tributaries that drain the valley slopes. The topographic and stratigraphic relationships between these units are shown in figure 3.4; detailed descriptions of each of the units can be found in Miller et al. (2001).

The oldest alluvial unit, Qa_1, is buried by younger deposits and has only been identified at the base of stratigraphic sections exposed along incised channels. At most sites, the surface of Qa_1 is characterized by a soil profile. The nature of the profile, and the degree of profile development, varies significantly between sites. Upstream of side-valley alluvial fan deposits, the buried soils are poorly developed, averaging approximately 15–45 centimeters in total thickness, and the profiles exhibit organic-rich A horizons that locally contain undecomposed plant debris. Radiocarbon dates of charcoal and organic-rich sediments from the buried Qa_1 soils range from 4180 ± 100 to 3450 ± 160 YBP (table 3.1). The relationships between Qa_1 and Qf_1 have not been clearly defined. At one location within the San Juan Creek basin, however, the two units were interpreted to interfinger, and

TABLE 3.1.

*Summary of radiocarbon dates obtained from delineated stratigraphic units
in study basins of the Toiyabe, Toquima, and Monitor Ranges*

Stratigraphic Unit Sample Number	Basin/Range	Material Dated	Determined Age (YBP)
Qa4			
SBR2A, RC3	Stoneberger Creek/Toquima	Charcoal	Modern
BR5b, RC16	Stoneberger Creek/Toquima	Charcoal	440 ± 50
CWJ1	Cottonwood Creek/Toiyabe	Charcoal	210 ± 50
CW5	Cottonwood Creek/Toiyabe	Charcoal	290 ± 50
Qa3			
BC4, RC21	Barley Creek/Monitor	Charcoal	1,200 ± 60
SJ3	San Juan Creek/Toiyabe	Bulk organic sediment	1,250 ± 50
SJJ2	San Juan Creek/Toiyabe	Charcoal	1,310 ± 50
Qa2			
BC3-A	Barley Creek/Monitor	Charcoal	1,920 ± 60
BC3-B	Barley Creek/Monitor	Charcoal	2,040 ± 60
CWJ3	Cottonwood Creek/Toiyabe	Charcoal	1,960 ± 50
SJJ4	San Juan Creek/Toiyabe	Charcoal	2,130 ± 50
SJ1	San Juan Creek/Toiyabe	Peat	2,270 ± 60
SJ5.75A	San Juan Creek/Toiyabe	Bulk organic sediment	2,580 ± 70
Qf2			
SB3, RC7	Stoneberger Creek/Toquima	Charcoal	2,110 ± 40
Qa1			
CW1	Cottonwood Creek/Toiyabe	Bulk organic sediment	3,450 ± 160
SJ7-B	San Juan Creek/Toiyabe	Charcoal + organic sediment	3,460 ± 70
SJ5	San Juan Creek/Toiyabe	Charcoal + organic sediment	4,180 ± 100

soils developed in the Qa_1 deposits were traced onto the surface of Qf_1 (fig. 3.4), a unit consisting of massive, clast-supported, pebble-sized gravels. The extensive nature of the soils observed on the Qa_1 surface suggests that valley floors within the upland watersheds were relatively stable from approximately 4200 to 3450 YBP. Local paleoclimatic data indicate that this period corresponds to the Neoglacial (approximately 3500 to 2600 YBP in this area) that was characterized by relatively cool and moist climatic conditions (Wigand 1987; Wigand and Rhode 2002; chapter 2).

Qa_1 is overlain by Qa_2, the most extensive alluvial unit within most watersheds (fig. 3.4). Sedimentologically, Qa_2 consists of both a fine- and coarse-grained facies. Most of the unit consists of the fine-grained facies that is characterized by silty clay loam textures that contain less than 15 percent gravel disseminated throughout the material. The unit also contains 1 to 10 centimeters thick, discontinuous layers of dark brown to black silt-sized sediment containing an abundance of charcoal. The less-prominent, coarse-grained facies is dominated by a clast-supported gravel with a fine sand to sandy loam matrix. This facies generally occurs as 0.5 to 2-meter-long lenses within the fine-grained facies.

Coarse-grained, clast-supported Qf_2 deposits interfinger with Qa_2 deposits in the vicinity of side-valley alluvial fans indicating that the two units were syndepositional. Almost without exception, Qf_2 deposits form the surface of side-valley fans within the upland watersheds and exceed thicknesses of several meters. Given the interfingering relationship between Qa_2 and Qf_2, it appears their deposition was produced by an intensive period of hillslope erosion during which time the hillslope sediments were shed onto the side-valley fans, causing significant fan aggradation. The sediments were then transported downstream, initiating several meters of aggradation on the valley floor. Deposition, based on seven radiocarbon dates from Qa_2 and Qf_2, occurred from approximately 2580 ±70 to 1900 ± 60 YBP (table 3.1). It is important to recognize that deposition appears to have been synchronous across the central Great Basin in spite of the fact that the upland watersheds are underlain by different rock types and characterized by different tectonic settings. This suggests that aggradation was initiated by a regional change in the environment. Given that the depositional episode predates European settlement, the most likely cause for aggradation was a change in climatic conditions.

Paleoecological studies have shown that a shift from relative wet conditions to warm and dry conditions began in the study area at about 2600 YBP and slightly later (about 2000 YBP) on a regional basis (Miller et al. 2001; Wigand and Rhode 2002; chapter 2; fig. 3.5). The drier and warmer conditions are thought to have decreased the extent of woodlands and to have increased the dominance of desert shrub vegetation throughout the northern Great Basin (Wigand et al. 1995). Data from packrat middens demonstrate that the total number of upland plant taxa locally decreased by 50 percent or more within and adjacent to the study basins (Miller et al. 2001; chapter 2). Interestingly, the onset of the drought coincides with the approximate timing of fan and valley floor aggradation. Thus, it is inferred that aggradation was associated with a change from moister to drier conditions and a significant change in the nature of the upland vegetation.

On a theoretical basis, the Langbein-Schumm model presented in figure 3.1 suggests that fluvial systems in the region may be particularly sensitive to climatic change. Although the effective precipitation for the area during the Neoglacial cannot be accurately determined, it is likely to have been on the order of 300 to 700 millimeters, depending on the local elevation.

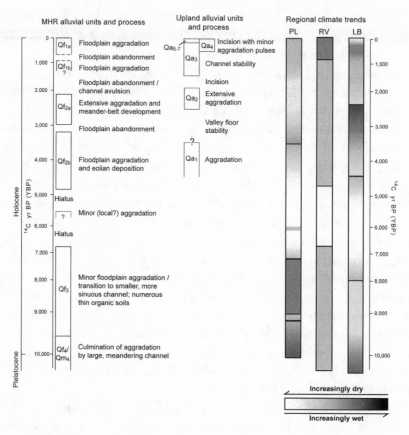

FIGURE 3.5. Comparison of middle Humboldt River and Upland alluvial chronologies with regional climatic trends. General fluvial processes related to each alluvial unit are indicated. The trends in regional climate are generalizations of data and interpretations presented in Benson et al. (2002) for Pyramid Lake (PL); in Thompson (1992) for Ruby Valley, Nev. (RV); and in Madsen et al. (2001) for Lake Bonneville Basin (LB).

This places the upland watersheds near the peak of the sediment yield curve (fig. 3.1) implying that the terrain is highly susceptible to climatic disturbances and that drier conditions would lead to increased sediment yields.

The argument that warmer and drier conditions lead to threshold-crossing events is also found in the geomorphic literature for a number of arid to semiarid regions (Bull and Schick 1979; Wells et al. 1987; Bull 1991; Ritter et al. 2000). Bull and Schick (1979), for example, presented a process-response model in which decreasing effective precipitation in southern Israel led to a loss of upland vegetation, an increase in the erodibility of the slope materials, and a change in hillslope hydrology. The net effect was a significant change in sediment loads to alluvial fans, and, consequently, aggradation. Wells et al. (1987) modified the Bull and Schick model by arguing that decreases in the infiltration capacity of hillslope soils also could result in increased runoff. In the case of the Mojave Desert where their study was conducted, they proposed that increased rates of hillslope pedogenesis, and decreased infiltration capacities of the soils, were produced by the deposition of fine-grained, eolian sediment and salt derived from the erosion of pluvial lake beds. The result was enhanced runoff from the less-permeable hillslopes that, when combined with increased sediment availability associated with a reduction in vegetation density, led to fan aggradation in spite of increasing aridity.

The upland watersheds in the central Great Basin appear to have been similarly affected by this complex interplay among changing climatic conditions, vegetation composition and density, hillslope hydrology, and sediment yields. In this case, the stratigraphic and paleoecological data suggest that a change in the type and density of the hillslope vegetation caused by a shift to drier, warmer conditions increased runoff and hillslope erosion. This change in sediment and water production initiated an episode of side-valley alluvial fan building and aggradation within the valley bottoms represented by Qa_2-Qf_2 deposits. This hypothesis is supported by the occurrence of multiple, thin but distinct layers of charcoal in Qa_2 deposits, implying that numerous wildfires may have occurred during its deposition. Similar layers are absent in the younger deposits, suggesting that these fires may have played a significant role in increasing hillslope sediment availability and runoff at the time of Qa_2 deposition. Although the relationships between fire frequency and climate are complex, it is reasonable to assume that the preservation of charcoal in Qa_2 is the result of more frequent fires, relatively dry conditions, and rapid aggradation.

The radiocarbon data indicate that deposition of Qa_2 had ceased by approximately 1900 YBP (table 3.1). Radiocarbon dates from the base of Qa_3 have shown that deposition began at approximately 1200–1300 YBP (table 3.1). Qa_3 occurs as thin (less than 1.5 meters) strath terrace deposits of generally uniform thickness inset into both Qa_2-Qf_2 materials, and its areal extent is highly variable along the valley floor of any given watershed. Along many reaches, it is absent. Deposition of Qa_3 during lateral channel migration appears to have continued until the most recent phase of channel incision was initiated, which, as will be discussed in more detail below, is highly variable both within and between basins. The occurrence of Qa_3 as a strath terrace, combined with its limited distribution, suggests that Qa_3

deposition was associated with a relatively stable channel system. Channel stability is supported by the sedimentology of the fine-grained facies contained in Qa_3. With the exception of the charcoal layers, Qa_3 is similar to that of Qa_2. The similarities presumably result from the reworking and redeposition of Qa_2 materials during lateral channel migration and strath terrace formation, with the limited introduction of new sediments from the hillslope. Climatically, Qa_3 deposition appears to correspond to a period of increased summer rainfall that occurred from approximately 1400 to 1000 YBP (Wigand and Rhode 2002).

From 1900 to 1300 YBP there is no evidence of significant aggradation within any of the study basins. Given that Qa_3 is inset into the older valley-fill deposits, this six-hundred-year interval appears to have been a period of minor channel incision with limited influx of materials from the hillslope. Paleoclimatic data indicate that the drought that began in this area at about 2600 YBP (chapter 2) intensified and expanded at about 2000 to 1900 YBP. Thus, it is possible that a further reduction in moisture led to a threshold-crossing event in which hillslope runoff was no longer capable of eroding slope materials. Such a decrease in runoff and sediment supply could lead to axial channel incision. Bull (1991) argues, however, that a change from aggradational to degradation processes does not necessarily require a change in climate. Rather, a reduction in sediment yields can result from the depletion of erodible materials stored on the hillslopes. Application of this model to the upland watersheds in the central Great Basin suggests that the fine-grained sediment that dominates Qa_2 deposits were preferentially removed from the slopes, producing the coarse-grained surface lags and gravelly A-horizons that are observed in many upland areas. This argument is supported by a comparison of unit sedimentology. Qa_2 tends to be finer grained than subsequently deposited stratigraphic units including Qa_3–Qa_7. The hillslope winnowing process that is envisioned here would have increased the infiltration capacities of the soils, reduced runoff from the slopes, and decreased sediment loads to the side-valley fans and axial channel systems. Although runoff from the hillslopes may have decreased, the sediment deficient nature of the flows within the axial channels would presumably have initiated channel incision.

Regardless of the exact mechanism that initiated channel incision at approximately 1900 YBP, it appears certain that a shift from moister to drier conditions resulted in two opposing geomorphic responses. The initial response was massive hillslope erosion accompanied by side-valley fan and valley floor aggradation. The final response was hillslope and alluvial fan stabilization, which was accompanied by minor channel incision. Moreover, it indicates that changes in channel processes may be initiated by internal adjustments within the watershed, in this case, a depletion of erodible hillslope sediment. In other words, the latter response was a by-product of the initial response.

The dominant process operating within the upland stream systems today is channel incision. In fact, nearly all of the stream systems that have been examined in the Toiyabe, Toquima, and Monitor Ranges exhibit evidence of recent channel-bed degradation. The most conclusive and widespread evidence for incision are the multiple terrace and terrace deposits (including Qa_5–Qa_7) that are preserved along the channel networks. These deposits are inset below the valley floor (and below Qa_3 where it is present). The maximum ages of these units have been estimated by dendrochronological dating of willows growing on their surface. Although there is significant variability in the age of individual trees, Qa_6 and Qa_7 correspond to flood events in 1983 and 1995 (Chambers et al. 1998). The ages of willow stands on Qa_5 are more variable, but cluster in the mid- to late 1970s when several high spring flows occurred (Miller et al. 2001). As expected, current channel incision is driven by moderate- to low-frequency, high-magnitude events.

The timing of this more recent phase of channel entrenchment has not been accurately determined. However, channel fill deposits associated with Qa_4 have been dated between the present and 440 ± 50 YBP (table 3.1), suggesting that at least localized entrenchment had began by about 450 YBP, perhaps in response to the Little Ice Age.

Qa_4 channel fill deposits are spatially limited to reaches located immediately upstream of side-valley alluvial fans that extended across the valley floor during the deposition of Qa_2-Qf_2. Locally, the channel fill deposits can be traced onto the valley floor where they occur as localized lobes of clast-supported gravels. The deposition of Qa_4 is interpreted to represent the rapid aggradation of low-gradient stream reaches during major flood events, a process that forces water onto the surface of the valley floor. These overbank flows subsequently produce new channels entrenched into the valley fill. The process, therefore, represents a form of channel avulsion that was observed within the upland watersheds as recently as 1995.

The localized nature of the Qa_4 paleochannel fills implies that the initial phase of incision may not have begun simultaneously along the stream systems. Rather, the limited distribution of Qa_4 paleochannel fills suggests that degradation was characterized by the formation of discontinuous zones of entrenchment. This is consistent with the current distribution of entrenched valley floors in a number of watersheds in the area (e.g., Indian Creek described in chapter 5) in which the current valley is characterized by unentrenched and entrenched reaches.

The factors driving the most recent episode of incision may include both changes in climate and a host of anthropogenic disturbances including livestock grazing, road construction, and water diversions. Analysis of the primary causes of entrenchment is currently underway in several drainage basins, but several factors hinder the assessment of channel entrenchment: (1) there is a lack of understanding of the types and magnitudes of land-use change that have occurred in upland watersheds; (2) there is an inability to precisely constrain the timing of incision during the past 150 years and their relations to historic shifts in climate; and (3) the record of incision is complex, presumably reflecting the influence of both internal and external controlling variables. What is important to recognize, however, is that the evolutionary history of the upland watersheds in the central Great Basin during the late Holocene appears to have produced hillslopes with only a limited amount of debris that can be removed from the sediment reservoirs. Thus, for the past 1,900 years, the axial channels have alternated between periods of stability (Qa_3 deposition) and channel incision; the primary response to environmental change has consistently been channel-bed degradation. For example, Germanoski and Miller (1995) examined the geomorphic impacts of a wildfire in 1981 on the Crow Canyon watershed, located within the Toiyabe Range near Austin, Nevada. The fire resulted in the complete destruction of all ground and tree foliage, leaving only charred trunks and branches. They found that hillslope erosion following the fire was limited, in spite of the fact that the largest precipitation and runoff events of the past century occurred in 1983, just two years after the fire occurred. The primary geomorphic response to the event was axial channel entrenchment that reached as much as 3.9 meters in depth within headwater areas. Although the Crow Canyon basin is underlain by coarse-grained granodiorites and quartz monzonites that would not be expected to produce large quantities of fine sediment, the data collected by Germanoski and Miller (1995) demonstrate that only limited hillslope erosion can be initiated in many upland watersheds of central Nevada under the current hydrologic regime. When combined with the previously presented data, it leads to the conclusion that any disturbance that leads to increased runoff within the study area, regardless of whether it is due to natural or anthropogenic processes, is likely to produce entrenched axial channels.

Influence of Late Holocene Landforms on Modern Channel Dynamics

As mentioned above, many, if not most, of the streams within the upland watersheds of the central Great Basin are currently undergoing a phase of channel-bed degradation. Detailed mapping of the depths of incision in ten basins in three ranges (the Toiyabe, Toquima, and Monitor) demonstrates that the entrenchment has not proceeded uniformly along channels, nor does it vary systematically in magnitude along the riparian corridor (chapter 4). Rather, it is highly variable and corresponds to the distribution of large, side-valley alluvial fans.

During the deposition of Qa_2-Qf_2, many of these side-valley alluvial fans appear, on the basis of radial fan profiles, to have blocked the downvalley movement of water and sediment (fig. 3.3). Ultimately, the distal-most segments of the fans were breached, reintegrating the axial drainage system that is present today. Stratigraphic and sedimentological data collected immediately upstream of the fans show that the valley fill in these areas is highly complex, consisting in many cases of crosscutting paleochannels and gravel lobes associated with Qa_4 that overlie or are inset in Qa_2 and Qa_3 deposits. There is no evidence to suggest that lacustrine conditions existed in these locations, although wetland (meadow complexes) presently occur upstream of many of the side-valley fans. Based on the lack of evidence of ponded flow, blockage of the axial channel system appears to have been temporary. Nonetheless, the side-valley fans enhanced upstream deposition by reducing local channel gradients and creating a bottleneck effect that reduced local flow velocities (fig. 3.6A). Longitudinal channel bed and valley floor surveys demonstrate that deposition of the Qf_2 deposits, combined with enhance aggradation upstream of the fans, led to a "stepped" profile in which significant changes in channel and valley floor elevations occur over a short downvalley distance (fig. 3.6B). Thus, the modern stream system is not characterized by a "smooth" concave-up longitudinal profile commonly associated with channels in equilibrium. Rather, the profile consists of a series of concave-up segments that terminate downstream at a side-valley fan that creates a "step" in the channel bed. Given that the disruption in channel-bed gradients are associated with the deposition of Qf_2, the discontinuities in gradient have been in existence for at least the past 1,900 years and are the direct result of hillslope and fluvial processes that occurred approximately two thousand years ago.

The slow rate at which these discontinuities in gradient have been re-

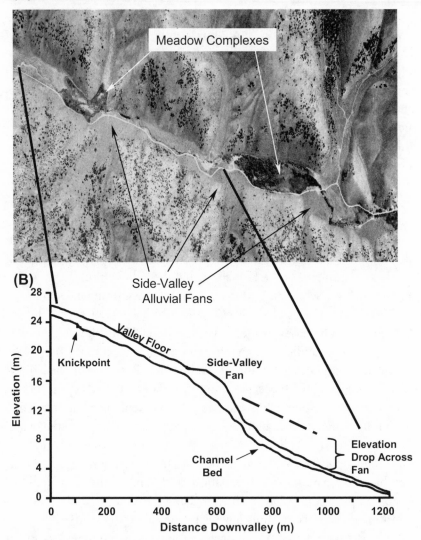

FIGURE 3.6. (A) Aerial photograph of side-valley fans in Kingston Canyon showing variations in valley morphology and vegetation types along the valley floor; and (B) longitudinal profile of the channel bed and valley floor showing the stepped nature of the valley topography. (Figure modified from Miller et al. 2001.)

moved from the streams is presumably related to two factors. First, the side-valley fans are breached at the lowest point in the valley topography that occurs along the contact between the fan deposits and the opposing hill-slope. Because of the proximity of the stream to hillslopes, resistant bedrock outcrops are common along the channel bed. Second, sediments eroded from the fan deposits tend to be coarse-grained and more poorly sorted than the materials found either upstream or downstream of the side-valley fans. Entrainment of these materials requires rare, high-magnitude flood events, thus reducing the rate at which incision can occur. When downcutting within the reach does occur, incision is propagated upstream, commonly in the form of a migrating knickpoint. Clearly, not all of the side-valley fans create discontinuities in the longitudinal profiles of the channel, but where they occur, they act as local base-level controls that dictate the magnitude of incision that transpires along upstream reaches.

Holocene Geomorphic Record of the Humboldt River Valley

Geographic and Geomorphic Setting

The Humboldt River is the largest single drainage system in the Great Basin. Its watershed drains approximately 43,000 square kilometers, including most of the upland watersheds previously described (fig. 3.2). The area of the watershed contributing flow to the reach of the Humboldt River examined in this study increases from approximately 23,000 square kilometers to nearly 30,000 square kilometers as it is joined by various tributaries. The Reese River, which is the axial drainage for several of the upland watersheds described earlier, is the largest of these (fig. 3.2).

As a consequence of the river's generally east-west traverse across the structural grain of the Basin and Range, and the alignment of major tributary valleys along a northeast-southwest trend, the Humboldt River basin is highly irregular in shape. The area drained by the Humboldt River is one of the driest regions in the intermountain West, and the river is correspondingly characterized by low streamflow in relation to the size of its watershed. The average annual discharge of the Humboldt River study reach is approximately 11 cubic meters per second (ninety-seven-year average of streamflow at the Humboldt River gage near Palisade, Nev.; USGS gage no. 10322500) and it occasionally drops to nearly zero flow in particularly dry years.

The Humboldt River study reach possesses a low gradient (valley slope about 0.004; channel slope about 0.0013) and the channel follows a tortuous course (sinuosity about 3.0) across an alluvial floodplain approximately 7 kilometers wide. The active floodplain flanking the river ranges from 1 to 4 kilometers in width, and the remaining valley bottom is composed of a complex mosaic of late Quaternary fluvial deposits. These alluvial deposits have recorded the geologic and geomorphic history of the basin, including the regionally integrated influence of Holocene climatic changes. Comparison of the fluvial history delineated for the upland watersheds with the history of this major regional drainage system provides an opportunity to evaluate the influence of scale on the response of fluvial systems to the same general external controls.

Methodology

Stratigraphic studies, combined with geologic mapping, along a 40-kilometer reach of the broad Humboldt River floodplain near Battle Mountain, Nevada (subsequently referred to as the middle Humboldt River, or MHR), have led to the development of an alluvial chronology that spans the Holocene and latest Pleistocene (House et al. 2000, 2001; Ramelli et al. 2001). The geologic maps delineate multiple generations of terrace deposits as well as complex sequences of coeval meander-belts. In general, terrace strata are the most significant alluvial units for developing a chronology. The number and characteristics of associated meander-belts provide a perspective on fluvial dynamics associated with the formation of the terraces.

The alluvial chronology is based on prominent unconformities in vertical section, and crosscutting and inset relations in planimetric (map view) section. Age control on the stratigraphy and surficial geology spans the last twelve thousand years and is based on (1) twenty-six radiometric dates from various types of organic detritus collected from alluvial deposits, including gastropod shells, charcoal, and organic-rich fluvial mud; (2) the stratigraphic position of the 7,600-year-old (about 6800 YBP) Mazama tephra (Zdanowicz et al. 1999); (3) prehistorical and historical artifacts; and (4) historical archives. Specific details about sample ages, materials, and locations are provided in tabular form on the geologic maps by House et al. (2000, 2001) and Ramelli et al. (2001).

Stratigraphic and geomorphic data collected from the Humboldt River's floodplain combine to provide an elaborate record of the influence of latest Pleistocene and Holocene climate change on the river's hydrologic regime. More specifically, alluvial strata, and the relations between individual stratigraphic units, exposed in gravel pits and cutbanks along the river and its major floodplain tributaries indicate that multiple episodes of floodplain aggradation and meander-belt development are separated by variably long periods of fluvial incision or relative inactivity (stability and nondeposition).

Overview of the Late Pleistocene and Holocene History of the Middle Humboldt River

The Humboldt River valley contains a complex assemblage of abandoned and active floodplain and terrace surfaces of low relief interspersed with topographically complex, active, and abandoned meander-belts (fig. 3.7). The various mappable geologic units have complex planimetric and cross-sectional relationships that represent episodes of floodplain aggradation, abandonment, and channel incision that are likely to be influenced by external factors, climatic variability in particular (fig. 3.8). Additionally, the Humboldt River traverses a tectonically active region and is therefore susceptible to change associated with tectonic processes.

The geomorphic history of the middle Humboldt River (MHR) is summarized below. Specific stratigraphic units are described according to map units used in House et al. (2000, 2001) and Ramelli et al. (2001). Unit nomenclature presented on their surficial geologic maps is different from that previously put forth here to describe the stratigraphy within the upland watersheds and published in Miller et al. (2001). Figure 3.5 provides for a direct comparison of the age of each stratigraphic unit delineated in the two study areas and can be used to clarify questions concerning unit nomenclature.

Along the MHR, the late Pleistocene culminated with an episode of fluvial aggradation by a large meandering river. The time at which this aggradational episode began has yet to be constrained. It continued until approximately 9500 YBP and led to the formation of units Qm_4 (meander-belt complex) and Qf_4 (coeval floodplain). Unit Qm_4 includes a particularly conspicuous series of overlapping megameanders that are much larger than the modern and mid-late Holocene meanders of the Humboldt River (fig. 3.7). Unit Qf_4 includes the coeval, flanking terrace deposits that are locally preserved on the outer margins of the valley floor. Together, these units represent the culmination of aggressive deposition and meander-belt migration by the Humboldt River during an episode of greatly enhanced stream flow and sediment discharge.

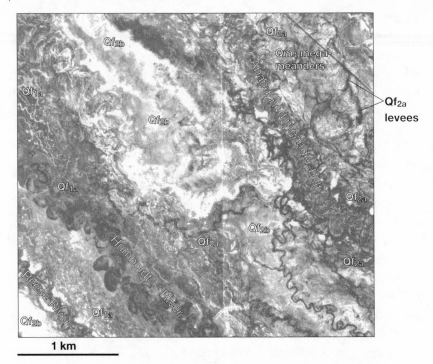

FIGURE 3.7. Aerial photograph of the Humboldt River floodplain located approximately 5 kilometers north-northwest of Battle Mountain, Nev. (north is to the top). Alluvial stratigraphic units and other geologic features described in text are indicated (generalized from House et al. 2001). Note that Qm_4 megameander-belt located in the northeast corner of the photo is largely buried by Qf_{2a} flood-plain sediments. The persistence of the older meander patterns at the surface is presumably due to the influence of coarse gravels on groundwater levels and related vegetation patterns. The meander-belt in the northeast corner of the photo was abruptly abandoned by the MHR (middle Humboldt River) approximately 2100 YBP. The Reese River (extreme southwest corner) occupies a more recently abandoned MHR meander-belt.

The aggradation and meander-belt activity that ended in the latest Pleistocene was followed by a period of episodic, relatively minor fine-grained floodplain aggradation, valley-bottom stability, and marsh deposition in the early Holocene. This interval is represented by unit Qf_3 and spanned approximately 9500–6800 YBP. Unit Qf_3 is characterized by dark, organic-rich floodplain mud. Locally, several thin layers of black mud are evident. The nature of this unit indicates generally wet conditions, but with decreased stream flow relative to the preceding interval that was characterized by the overlapping megameanders.

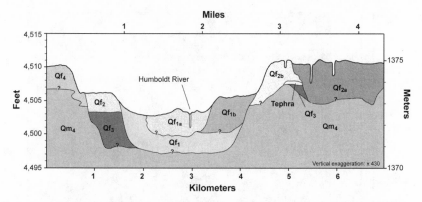

FIGURE 3.8. Generalized cross section of the alluvial stratigraphy of the middle Humboldt River (MHR) floodplain (simplified from House et al. 2001). In this rendition, meander-belt deposits have been combined with their coeval floodplain deposits. See text for explanation of units. Note the large quantity of vertical exaggeration used in the diagram.

The middle Holocene record of the Humboldt River (approximately 6800–4800 YBP) is characterized by the influence of relative aridity and significantly reduced stream flow. The most compelling evidence for these conditions is a widespread, major unconformity in the alluvial stratigraphic record indicating greatly reduced deposition by the river and increased deposition of windblown silt. There is some local evidence for a minimal amount of floodplain aggradation ending at approximately 5500 YBP (Unit Qf_{2c}), but overall, the record is one of inactivity. The beginning of this relatively dry interval is delineated by the contact of Mazama tephra (about 6800 YBP) with the underlying dark muds of Qf_3 (fig. 3.9). In some exposures, floodplain deposits less than 30 centimeters above the tephra bed postdate it by up to 4,000 years (fig. 3.10).

The middle Holocene dry interval was followed by a gradual increase in fluvial activity. A period of widespread floodplain aggradation and accumulation of eolian silt occurred along the Humboldt River between approximately 4800 and 3300 YBP. This episode is represented by unit Qf_{2b} which overlies unit Qf_3. Qf_{2b} deposition is associated with the formation of a prominent floodplain terrace (fig. 3.7). Fluvial sediments in unit Qf_{2b} are distinctly less organic-rich and are interbedded with, and overlain by, considerably more eolian silt than the next younger unit—Qf_{2a}. In many exposures, the lower part of the Qf_{2b} alluvial package contains a large amount of reworked Mazama tephra despite the fact that the aggradation apparently commenced nearly two thousand years after the tephra was deposited across the watershed.

FIGURE 3.9. Photograph of 1.5-meter-high high bank exposure of floodplain alluvium along Rock Creek, a major tributary to the middle Humboldt River. Here, the Rock Creek channel occupies an abandoned MHR meander-belt. This exposure clearly indicates the striking contrast at the contact between the organic-rich mud deposits typical of early Holocene unit Qf_3, and the overlying volcanic ash (the Mazama tephra) and mixed overbank eolian sediments of unit Qf_{2b}.

During the interval of approximately 3300 to 3000 YBP, the Humboldt River progressively abandoned and laterally eroded Qf_{2b} floodplain deposits. Also during this interval, eolian silt continued to accumulate on the abandoned Qf_{2b} floodplain surface.

A period of extensive meander-belt migration and floodplain aggradation occurred between approximately 3000 and 2100 YBP. The period is

FIGURE 3.10. Gravel-pit exposure of MHR channel and floodplain alluvium. At this site, a thick sequence of cross-bedded late Pleistocene meander-belt gravels are immediately overlain by thick Qf_3 deposits capped with a clean bed of Mazama tephra. The tephra is overlain by eolian and fluvial silt and sand corresponding to unit Qf_{2b}. Organic-rich floodplain alluvium of unit Qf_{2a} overlies Qf_{2b} here along an erosional unconformity. The geologist in the photo is approximately 2 meters tall.

represented stratigraphically by unit Qf_{2a}, which is inset against and locally overlies Qf_{2b}. Unit Qf_{2a} is associated with a distinctly sediment-laden meandering channel. It includes channel-flanking levee complexes with 2–3 meters of relief adjacent to a series of overlapping meander-belts that indicate rapid channel migration (fig. 3.7). Rapid channel and overbank aggradation at this time resulted in an elevated channel-levee complex that was entirely abandoned via avulsion by the Humboldt River at about 2100 YBP, the youngest radiometric age from the floodplain surface flanking the most recent abandoned meander-belt. Organic-rich sediments from the crests of similar extant levees associated with older Qf_{2a} meander-belts yield ages of approximately 2500 YBP, possibly indicating the time of maximum aggradation.

Abandonment of the youngest Qf_{2b} meander-belt resulted in a shift in the location of the main Humboldt River channel from the north side of the floodplain to the south side. This development was accompanied by channel incision and lateral erosion of older floodplain deposits between approximately 2100 and 1200 YBP. Some or most of this erosion could have been driven by channel relocation. Incision was followed by flood-plain aggradation from approximately 1200 to 700 YBP, which resulted in unit Qf_{1b}, a floodplain surface that is inset against and slightly below extant Qf_2 surfaces (fig. 3.7).

Partial floodplain abandonment and minor incision occurred from ap-proximately 700–550 YBP, and was followed by floodplain aggradation of unit Qf_{1a} from approximately 550 YBP to the present. Unit Qf_{1a} is inset against and slightly below Qf_{1b}. The youngest Qf_{1a} deposits are historical and either bury or contain nineteenth and twentieth century artifacts.

The most significant geomorphic events contained in the Holocene al-luvial chronology include the major hiatus in fluvial deposition between 6800 and 4800 YBP (which corresponds with drought conditions reported from various sites in the Great Basin; see discussion below); distinct peri-ods of floodplain aggradation and intermittent eolian deposition between about 4800 and 3300 YBP; and a period of lateral plantation, incision, and apparently rapid floodplain aggradation between approximately 3000 and 2100 YBP. The latter appears to be the most dynamic Holocene episode along the MHR on the basis of multiple, overlapping meander-belts and extensive levee complexes, and its termination is directly associated with a major shift in channel location on the broad floodplain.

Relation of the MHR Alluvial Chronology to Regional Paleoclimate Reconstructions

The Holocene paleoenvironmental history of the Great Basin has been studied in various degrees of detail at many different sites using different techniques and data types (see Grayson 1993 for a comprehensive review). Recently published compilations from the Pyramid Lake Basin (Benson et al. 2002) and the Lake Bonneville–Great Salt Lake Basin (Madsen et al. 2001) also provide useful insights into the paleoclimatic changes that af-fected the MHR because they record changes in climate that occurred at the western and eastern margins of the Great Basin, respectively. Another useful source of data comes from Ruby Marsh in Ruby Valley, Nevada (Thompson 1992). Ruby Valley is a closed basin in eastern Nevada that shares a divide with the eastern margin of the Humboldt River watershed.

Comparisons of these regional paleoclimate reconstructions indicate that, though similar, there is a general lack of synchroneity in terms of the onset and duration of climatic change across the Great Basin (chapter 2). Some difference is to be expected because the records are composites of several indicators and no single indicator is a precise measure of climatic regime. Moreover, actual changes are not likely to be synchronous across a large region. Major trends in regional climate reconstruction do, however, overlap, and it appears that significant episodes in the Holocene alluvial chronology of the MHR are closely related to major climatic trends (fig. 3.5).

All of the paleoclimatic reconstructions listed above indicate an early Holocene interval of relatively wet conditions. Although the records emphasize different degrees of variability within the interval, they all indicate that it spanned a period of time from at least 10,000 YBP to approximately 7000 YBP. The alluvial chronology of the MHR suggests that the river during this period was transformed from a large, meandering system to a relatively slowly aggrading system, but apparently with a moist floodplain environment (unit Qf_3). Qf_3 deposits are characteristically rich with organic mud, indicating persistent saturated conditions.

There is consistent regional evidence that the middle Holocene was characterized by prolonged drought of spatially variable length and degree of severity. Benson et al. (2002) note, for example, that the interval 7300–3500 YBP was dry overall but with episodes of wide variability. Within this interval, they describe the period of 6800–4900 YBP as one of likely "intense aridity," whereas, the period of 5700–3500 YBP as one in which drought significantly decreased the flow into Pyramid Lake from the Truckee River. The latter observation is consistent with other data that indicate closure of Lake Tahoe, a principal source of Truckee River flow (Grayson 1993). The record from Ruby Valley indicates severe drought persisting from approximately 6800 to 4500 YBP. Regional pollen and midden data collected from the northern Great Basin indicate that the extreme drought of the middle Holocene came to an end at approximately 5400 YBP, and that while temperatures remained warm, there was a gradual increase in precipitation from approximately 5400 to 4000 YBP (Wigand 1987; Wigand and Rhode 2002).

The middle Holocene alluvial record from the Humboldt River is largely one of nondeposition. With only one exception, the Humboldt chronology contains an apparent hiatus in fluvial deposition spanning the period of approximately 6800 to 4800 YBP with only a brief pulse of activity at about 5500 YBP. At other locations, a tremendous gap from two thousand up to four thousand years long is present in less than 0.50 meters of alluvial section with no obvious erosional unconformities (aside from the possibility of deflation by wind).

The late Holocene paleoenvironmental data show a generally higher degree of climatic variability within an overall wetter climate than the previous interval. For example, Benson et al. (2002) emphasize that by at least 3000 YBP, the Truckee River began again to contribute significant input to Pyramid Lake, and Thompson (1992) notes that the lake in Ruby Valley began to recover from the middle Holocene drought beginning as early as 4700 YBP, becoming substantially more wet by 4300 YBP. Similarly, Madsen et al. (2001) argue that the Great Salt Lake in the Bonneville Basin began to recover from extreme drought by 4400 YBP, eventually peaked at 3400 YBP, and remained relatively high for the subsequent 1,400 years. Madsen et al. (2001) also emphasize that the Bonneville record indicates a dramatic transition to moister and cooler conditions during the interval of 2950 to 2400 YBP.

The remaining part of the late Holocene is typically characterized as relatively wet, with a few exceptions. For example, Wigand and Rhode (2002) suggest on the basis of pollen and woodrat midden data that the climate in the Great Basin became warmer and drier between approximately 1900 and 1000 YBP, although local increases in summer rainfall may have occurred between about 1400 and 900 YBP. Benson et al. (2002) also suggest it was relatively dry between approximately 2500 and 1700 YBP, whereas other records seem to indicate the likelihood of relatively wet conditions in the same general area during this time (see summary by Grayson 1993, 223–226). Benson et al. (2002) also note the occurrence of two intense droughts in the interval between 1250 and 570 YBP. Similarly, Madsen et al. (2001) mentions the likelihood of drought conditions in the Bonneville Basin between 700 and 600 YBP. This late episode of regional drought was soon followed by the so-called Little Ice Age (approximately 500–150 YBP), which is generally considered to be associated with a distinctly cool and wet interval (e.g., Grayson 1993; Madsen et al. 2001).

The MHR unit Qf_{2b} corresponds with the transition from drought to a relatively wet climate between 4800 and 3300 YBP. Qf_{2b} is composed of a mixture of abundant reworked Mazama tephra, eolian silt, and floodplain silt and sand. It is possible that the Qf_{2b} aggradation reflects the delivery of large quantities of ash and reworked eolian sediments from tributary watersheds. It is notable that the Qf_{2b} deposits comprise the highest Holocene terrace in the valley bottom and thus represent a significant net aggradation event.

The Qf_{2a} floodplain deposits are both inset against and overlie truncated parts of the Qf_{2b} terrace. The aggradation of Qf_{2a} alluvium occurred between about 3000 and 2100 YBP. This unit is associated with a series of prominent overlapping meander-belts and is conspicuously rich with organic-rich floodplain mud indicative of persistently wet conditions. Deposition of Qf_{2a} alluvium coincides well with evidence for a distinctly wetter climate (particularly in relation to the Bonneville Basin record). This is consistent with the clear evidence for Qf_{2a} being associated with a relatively dynamic river.

The Qf_1 series of units correspond to the most recent late Holocene environmental changes. The onset of Qf_{1b} deposition is not well constrained, but it occurred sometime between 1000 and 2000 YBP. This floodplain was abandoned sometime between 1000 and approximately 700 YBP, which suggests a possible correspondence with the transition into the Little Ice Age.

Comparison of the Upland Watershed and MHR Chronologies

The comparison of alluvial stratigraphic records from the MHR and the upland watersheds is complicated by differences in resolution of the radiocarbon-dated chronologies and the uncertainties inherent in radiometric age dating of fluvial deposits. Nonetheless, there is a general temporal correspondence between discrete alluvial units preserved within the upland watersheds and the stratigraphy of the Humboldt River (fig. 3.5). The principal discrepancies are related to the onset and duration of each event and probably reflect the influence of system scale and location on the timing and rate of response to change.

The oldest alluvial unit identified in the upland watersheds is Qa_1. It represents a period of relatively slow aggradation during the Neoglacial. While Qa_1 deposition was occurring in the upland watersheds, floodplain aggradation characterized the Humboldt River, resulting in the middle- to late-aged Qf_{2b} deposits. Aggradation in both areas ceased at approximately 3500 YBP. Valley-floor stability in the uplands, and incision-floodplain abandonment along the Humboldt River, occurred for the next three hundred to five hundred years. Minor incision and valley-floor stability in the uplands also correspond to a period of incision and floodplain abandonment along the Humboldt River from approximately 1900 to 1300 YBP.

From 2500 to 2100 YBP, both areas experienced one of the most intense periods of aggradation of the late Holocene. However, floodplain aggradation along the Humboldt River appears to have been initiated approximately several hundred years before intense hillslope erosion and valley floor aggradation occurred within the upland drainage systems. Moreover, aggradation along the Humboldt River appears to have been associated with high runoff and sediment transport rates that led to extensive meander-belt migration, the construction of high-relief levees, and floodplain aggradation. This interpretation is consistent with the onset of aggradation during the Neoglacial characterized by cool, moist conditions. In contrast, valley floor aggradation from approximately 2500 to 1900 YBP in the uplands has been attributed in the above discussion to a shift from moister to drier climatic conditions associated with a drought that followed the Neoglacial. The seemingly contradictory conclusion that hillslope erosion and valley aggradation in the uplands was initiated by a shift from moister to drier conditions at 2500 YBP while the Humboldt River continued to aggrade as a result of abundant runoff until approximately 2100 YPB may be explained by (1) a longer lag time in the hydrologic and geomorphic responses to climate change in the significantly larger Humboldt River watershed as a result of greater variability in elevation, vegetation, geographic distribution, and geologic materials, and (2) enhanced variability in the timing and nature of the climatic changes that occurred across the larger Humboldt River basin, which extends into relatively high-elevation tablelands and mountains along the northeastern margin of the Great Basin. Differential timing in climatic change in areas of southeastern Oregon (near the perimeter of the Humboldt basin) relative to the central Great Basin have been noted in other types of paleoclimate data (Wigand and Rhode 2002).

The temporal discrepancy in the termination of the principal late Holocene aggradation events between the two areas (2100 versus 1900 YBP) is minimal given the temporal resolution of the alluvial chronologies. However, it may indicate that sediment deposited upon the floodplains of the Humboldt River was not directly derived from the upland watersheds examined in this study. Aggradation along the Humboldt River appears to have ceased while significant hillslope erosion and valley aggradation was occurring in the uplands. These differences are not surprising because sediments derived from the upland watersheds are delivered to alluvial plains along the mountain fronts where the debris may be stored either on alluvial fans, or within the valley fill of the intermountain basins before it is transported to the Humboldt River. For example, several of the upland watersheds examined in this study (including Big Creek, Cottonwood Creek, San Juan Creek, and Washington Creek) are located on the eastern flank of the Toiyabe Range. Sediments derived from these watersheds must be transported across extensive alluvial fans before entering the Reese River, one of the largest tributaries to the Humboldt. Preliminary alluvial chronologies constructed by D. Germanoski and J. Miller (unpublished data) have shown that aggradational and erosional events, constrained by the occurrence of tephra deposits of known age, are spatially variable along the Reese River. Moreover, some reaches appear to be prone to entrenchment during the Holocene, whereas others are more likely to experience aggradation creating lobate, fanlike features on the valley floor. Thus, the delivery of sediment from upland watersheds to the MHR is regulated by sediment transport and storage processes operating on alluvial fans and the valley floors of large tributary basins to the Humboldt River. As a result, variations in the nature and timing of erosional and depositional events between the two areas can be expected.

The controls on sediment transport processes also explain the current differences in geomorphic process observed with the two areas. The uplands are presently undergoing an episode of entrenchment, whereas the middle Humboldt River is undergoing a phase of floodplain aggradation (in areas where the channel has not been artificially straightened). The sediment deposited on the Humboldt floodplain is not, however, derived from this most recent period of upland erosion as most of the sediment is deposited on alluvial fans at the mouth of the basins (see, for example, Germanoski and Miller 1995).

The similarities described above between the chronologies developed for the upland watersheds and MHR suggest that major episodes of erosion and deposition are driven by climatic factors. The relations between geomorphic events and changes in climate are most readily apparent along upland streams, which are closer to zones of water and sediment production. Differences, particularly with regard to the exact timing and duration of the events, occur as a result of sediment transport, storage, and erosional processes that are governed in part by spatial scale.

Management and Restoration Implications

Age-constrained alluvial chronologies were developed for small, upland watersheds in the central Great Basin. Comparison to local and regional paleoclimatic records suggest that the initial response from approximately 2500 to 1900 YBP was extensive hillslope erosion, fan building, and valley aggradation. Since approximately 1900 YBP, the studied watersheds have been characterized by channel incision, or periods of valley floor stability and strath terrace formation associated with processes of lateral channel migration. The change in geomorphic processes at 1900 YBP appears to have resulted from the removal of fine-grained materials from the slopes that altered the complex interplay between infiltration, runoff, and sediment yields in such a way as to have decreased slope erosion. The significance of the observations to watershed management and restoration is that the basins are prone to channel entrenchment and will remain in this condition until fine-grained sediments accumulate within the hillslope sediment reservoirs, a process that may require thousands of years to occur (Bull 1991). Moreover, returning the watersheds to a presettlement condition, as is commonly proposed for many restoration projects (National Research Council 1992), returns them to an unstable state.

The magnitude of incision varies both between the upland basins and within any given basin. Differences in incision between basins is discussed in the following chapter. Variations in the magnitude of entrenchment within a basin is primarily related to deposition on side-valley alluvial fans from approximately 2500 to 1900 YBP. Thus, the geomorphic processes that led to aggradation on the side-valley fans more than 1,900 years ago are still having a significant effect on modern channel dynamics. The magnitude of these effects varies as a function of watershed sensitivity to disturbance as described in chapter 4. Where the influence of side-valley alluvial fans is significant, however, the fan deposits serve as local base-level controls that dictate the rate and extent of upstream incision. Given that the lowering of base level is caused by erosion of the fan deposits, it seems plausible that entrenchment can be slowed by controlling the rate and magnitude of channel-bed degradation where the streams traverse the toes of the side-valley fans. Implementation of grade-control measures at the toe of the side-valley fans will be aided by narrow valley widths and channel margins bordered by bedrock hillslopes and coarse-grained fan deposits, both of which limit lateral channel migration.

The MHR chronology spans the entire Holocene. The longer alluvial record, in comparison to the upland basins, reflects differences in sediment storage capacities. The broad Humboldt floodplain provides an immense repository for alluvium, whereas the relatively small valleys of the upland watersheds have a limited capacity for sediment storage. The MHR chronologies demonstrate that the Holocene record has been characterized by a complex history of changing geomorphic processes and landforms. The most significant events inferred from the alluvial chronology include (1) a major hiatus in fluvial deposition between 6800 and 4800 YBP, (2) a distinct period of alternating floodplain aggradation and intermittent eolian deposition between about 4800 and 3300 YBP, and (3) a period of lateral migration, incision, and rapid floodplain aggradation between approximately 3000 and 2100 YBP. On the basis of multiple, overlapping meander-belts, and the formation of extensive levee complexes, the latter episode appears to represent the most dynamic period during the Holocene along the MHR. In general, there is a temporal correspondence between discrete alluvial units preserved within the upland watersheds and the MHR. The observed similarities suggest that major episodes of erosion and deposition are driven by climatic factors. Differences with regard to the exact timing and duration of the events are presumably related to the influence of spatial scale on runoff regimes as well as sediment transport and storage dynamics.

Many researchers, land managers, and restorationists focus on the impacts of anthropogenic disturbances on channel processes and form. The geomorphic, stratigraphic, and paleoclimatic data presented here indicate that past changes in channel dynamics have been driven to a significant degree by minor climatic perturbations. It follows, then, that modern channel processes will also be affected by climate change. Excluding the influences of climate change on modern channel processes may lead to flawed management and restoration strategies. More specifically, management strategies should recognize that fluvial systems of the central Great Basin are not static, but adjust to rather modest changes in climate. Methods that attempt to maintain the channel in its current form over periods of decades (i.e., without allowing for changes in channel morphology associated with climatic variations) are doomed to failure.

Acknowledgments

The authors want to thank Robert Barr, Dorothea Richardson, and Karen Waltman for their assistance with data collection and manipulation. The manuscript was greatly improved through the helpful reviews provided by Dr. Dale Ritter, Dr. Scott Eaton, and Dr. Mark Lord. Interpretations and shortcomings of the chapter are those of the authors. Funding for this research was provided, in part, by the USDA Forest Service's Great Basin Ecosystem Management Project.

LITERATURE CITED

Arnell, N. 1989. *Human influences on hydrological behaviour: An international literature survey*. Technical Documents in Hydrology. Paris: UNESCO.

Balling, R. C., Jr., and S. G. Wells. 1990. Historic rainfall patterns and arroyo activity within the Zuni River drainage basin, New Mexico. *Annals of the Association of American Geographers* 80:603–617.

Benson, L., M. Kashgarian, R. Rye, S. Lund, F. Paillet, J. Smoot, C. Kester, S. Mensing, D. Meko, and S. Lindström. 2002. Holocene multidecadal and multi-centennial droughts affecting northern California and Nevada. *Quaternary Science Reviews* 21:659–682.

Birkeland, P. W. 1984. *Soils and geomorphology*. New York: Oxford University Press.

Bryan, K. 1928. Historic evidence on changes in the channel of the Rio Puerco, a tributary of the Rio Grande in New Mexico. *Journal of Geology* 361:265–282.

Bull, W. B. 1991. *Geomorphic response to climatic change*. New York: Oxford University Press.

Bull, W. B., and A. P. Schick. 1979. Impact of climatic change on an arid watershed, Nahal Yael, southern Israel. *Quaternary Research* 11:153–171.

Chambers, J. C., K. Farleigh, R. J. Tausch, J. R. Miller, D. Germanoski, D. Martin, and C. Nowak. 1998. Understanding long- and short-term changes in vegetation and geomorphic processes: The key to riparian restoration. Pp. 101–110 in *Rangeland management and water resources*, edited by D. F. Potts. Herndon, Va.: American Water Resources Association, and Society for Range Management.

Chatters, J. C., and K. A. Hoover. 1992. Response of the Columbia River fluvial system to Holocene climatic change. *Quaternary Research* 37:42–59.

Cooke, R. U., and R. W. Reeves. 1976. *Arroyos and environmental change in the American south-west*. Oxford, UK: Clarendon Press.

Dohrenwend, J. C., B. A. Schell, and B. C. Moring. 1992. *Reconnaissance photogeologic map of young faults in the Millett 1° × 2° quadrangle, Nevada*. U.S. Geological Survey, Map MF-2176.

Downs, P. W., and K. J. Gregory. 1993. The sensitivity of river channels in the landscape system. Pp. 15–30 in *Landscape sensitivity*, edited by D. S. G. Thomas and R. J. Allison. Chichester, UK: John Wiley & Sons.

Dutton, C. E. 1882. *Tertiary history of the Grand Canyon district*. U.S. Geological Survey Monograph 2.

Fuller, I. C., M. G. Macklin, J. Lewin, D. G. Passmore, and A. G. Wintle. 1998. River response to high-frequency climate oscillations in southern Europe over the past 200 k.y. *Geology* 26:275–278.

Germanoski, D., and J. R. Miller. 1995. Geomorphic responses to wildfire in an arid watershed, Crow Canyon, Nevada. *Physical Geography* 16:243–256.

Gerrard, J., 1993. Soil geomorphology, present dilemmas and future challenges. *Geomorphology* 7:61–84.

Gifford, G. F., and R. H. Hawkins. 1978. Hydrologic impact of grazing on infiltration: A critical review. *Water Resources Research* 14:305–313.

Grayson, K. K. 1993. *The desert's past: A natural prehistory of the Great Basin*. Washington, D.C.: Smithsonian Institution Press.

Hereford, R. 1984. Climate and ephemeral-stream processes: Twentieth-century geomorphology and alluvial stratigraphy of the Little Colorado River, Arizona. *Geological Society of America Bulletin* 95:654–668.

———. 1986. Modern alluvial history of the Paria River drainage basin, southern Utah. *Quaternary Research* 25:293–311.

House, P. K., A. R. Ramelli, and C. T. Wrucke. 2001. *Geologic map of the Battle Mountain Quadrangle, Lander County, Nevada*. Nevada Bureau of Mines and Geology Map 130, 1:24,000 scale. Reno, NV: Nevada Bureau of Mines and Geology.

House, P. K., A. R. Ramelli, C. T. Wrucke, and D. A. John. 2000. *Geologic map of the Argenta Quadrangle, Nevada*. Nevada Bureau of Mines and Geology Open-File Map 2000-7, 1:24,000 scale. Reno, NV: Nevada Bureau of Mines and Geology.

Jacobs, L. 1991. *Waste of the west: Public lands ranching*. Published by author, P.O. Box 5784, Tucson, Arizona.

Karlstrom, E. T. 1983. Soils and geomorphology of northern Black Mesa. Pp. 313–342 in *Excavations on Black Mesa, 1981, a descriptive report*, edited by F. E. Smiley, D. L. Nichols, and P. E. Andrew. Research Paper 36. Southern Illinois University, Carbondale, Center for Archaeological Investigations.

Karlstrom, E. T., and T. N. V. Karlstrom. 1986. Late Quaternary alluvial stratigraphy and soils of the Black Mesa–Little Colorado River areas, northeastern Arizona. Pp. 71–92 in *Geology of Central and Northern Arizona: Field trip guidebook*, edited by J. D. Nations, C. M. Conway, and G. A. Swann. Geological Society of America Rocky Mountain Section Meeting, Flagstaff, Arizona.

Karlstrom, T. N. V., G. J. Gumerman, and R. C. Euler. 1974. Paleo-environmental and cultural changes in the Black Mesa region northeastern Arizona. Pp. 768–792 in *The geology of northern Arizona with notes on archaeology and paleoclimate*, edited by T. N. V. Karlstrom, G. A. Swann, and R. A. Eastwood. Geological Society of America Rocky Mountain Section Meeting, Flagstaff, Arizona.

Kleinhampl, F. J., and J. I. Ziony. 1985. *Geology of northern Nye County, Nevada*. Nevada Bureau of Mines and Geology Bulletin 99A, Reno, NV.

Knighton, D. 1998. *Fluvial forms and processes: A new perspective*. London: Arnold.

Knopf, F. L., R. R. Johnson, T. Rich, F. B. Samson, and R. C. Szaro. 1988. Conservation of riparian ecosystems in the United States. *Wilson Bulletin* 100:272–284.

Knox, J. C. 1984. Fluvial responses to small scale climatic changes. Pp. 318–342 in *Developments and applications of geomorphology*, edited by J. C. Costa and P. J. Fleisher. Berlin: Springer-Verlag.

Kochel, R. C., D. F. Ritter, and J. R. Miller. 1997. Geomorphic responses to minor cyclic changes, San Diego County, California. *Geomorphology* 19:277–302.

Langbein, W. B., and S. A. Schumm. 1958. Yield of sediment in relation to mean annual precipitation. *Transactions of the American Geophysical Union* 39:1076–1084.

Macklin, M. G., and J. Lewin. 1993. Holocene river alluviation in Britain. *Zeitschrift für Geomorphologie*, Supplemental Bulletin 88:109–122.

Madsen, D. B., D. Rhode, D. K. Grayson, J. M. Broughton, S. D. Livingston, J. Hunt, J. Quade, D. N. Schmitt, and M. W. Shaver III. 2001. Late Quaternary environmental change in the Bonneville Basin, western USA. *Palaeogeography, Palaeoclimatology, Palaeoecology* 167:243–271.

McFadden, L. D., and J. R. McAuliffe. 1997. Lithologically influenced geomorphic responses to Holocene climate changes in the southern Colorado Plateau, Arizona: A soil-geomorphic and ecological perspective. *Geomorphology* 19:303–332.

Miller, J. R., D. Germanoski, K. Waltman, R. Tausch, and J. Chambers. 2001. Influence of late Holocene hillslope processes and landforms on modern channel dynamics in upland watersheds of central Nevada. *Geomorphology* 38:373–391.

Miller, J. R., and R. C. Kochel. 1999. Review of Holocene hillslope, piedmont, and arroyo system evolution in southwestern United States: Implications to climate-induced landscape modifications in southern California. *Proceedings of the Southern California Climate Symposium: Trends and extremes of the past 2,000 years*, pages 139–192. Technical Reports. Natural History Museum of Los Angeles County, Los Angeles.

National Research Council. 1992. *Restoration of aquatic ecosystems*. Washington, D.C.: National Academy Press.

Newson, M., and J. Lewin. 1991. Climatic change, river flow extremes and fluvial erosion-scenarios for England and Wales. *Progress in Physical Geography* 15:1–17.

Ramelli, A. R., P. K. House, C. T. Wrucke, and D. A. John. 2001. *Geologic map of the Stony Point Quadrangle, Lander County, Nevada*. Nevada Bureau of Mines and Geology Map 131, 1:24,000 scale. Reno, NV: Nevada Bureau of Mines and Geology.

Ritter, D. F., R. C. Kochel, and J. R. Miller. 2002. *Process geomorphology*. 4th ed. New York: McGraw-Hill.

Ritter, J. B., J. R. Miller, and J. Husek-Wulforst. 2000. Environmental controls on alluvial fan evolution in Buena Vista valley, north central Nevada, during the late Quaternary time. *Geomorphology* 36:63–87.

Rumsby, B. T., and M. G. Macklin. 1994. Channel and floodplain response to recent abrupt climate change: The Tyne Basin, northern England. *Earth Surface Processes and Landforms* 19:499–515.

Schumm, S. A., and G. R. Brakenridge. 1987. River response. Pp. 221–240, in *North America and adjacent oceans during the last deglaciation*, vol. K-3 of the *Geology of North America* series, edited by W. F. Ruddiman and H. E. Wright Jr. Vol. K-3. Boulder, Colo.: Geological Society of America.

Schumm, S. A., and R. W. Lichty. 1965. Time, space, and causality in geomorphology. *American Journal of Science* 263:110–119.

Singer, M. J., and P. Janitzky. 1986. *Field and laboratory procedures used in a soil chronosequence study*. U.S. Geological Survey Bulletin 1648. Washington, D.C.: U.S. Government Printing Office.

Soil Conservation Service. 1981. *Examination and description of soils in the field, in soil survey manual*. Washington, D.C.: U.S. Government Printing Office.

Stewart, J. H., and J. E. Carlson. 1976. *Geologic map of north-central Nevada*. Nevada Bureau of Mines and Geology Map 50, 1:250,000 scale. Reno, NV: Nevada Bureau of Mines and Geology.

Thompson, R. S. 1992. Late Quaternary environments in Ruby Valley, Nevada. *Quaternary Research* 37:1–15.

Trimble, S. W., and A. C. Mendel. 1995. The cow as a geomorphic agent: A critical review. *Geomorphology* 13:233–253.

Wallace, R. E. 1987. Grouping and migration of surface faulting and variations in slip rates on faults in the Great Basin Province. *Bulletin of the Seismological Society of America* 77:868–876.

Wells, S. G., L. D. McFadden, and J. C. Dohrenwend. 1987. Influence of late Quaternary climatic changes on geomorphic and pedogenic processes on a desert piedmont, eastern Mojave Desert, California. *Quaternary Research* 27:130–146.

Wigand, P. E. 1987. Diamond Pond, Harney County, Oregon: Vegetation history and water table in the eastern Oregon desert. *Great Basin Naturalist* 47:427–458.

Wigand, P. E., M. L. Hempill, S. S. Sharpe, and S. Patra. 1995. Great Basin semi-arid woodland dynamics during the late Quaternary. Pp. 51–70 in *Proceedings: Climate change in the four corners and adjacent regions: Implications for environmental restoration and land-use planning*, edited by W. J. Waugh; Sept. 12–14, 1994; Mesa State College, Grand Junction, Colo.: U.S. Department of Energy.

Wigand, P. E., and D. Rhode. 2002. Great Basin vegetation history and aquatic systems: The last 150,000 years. Pp. 309–367 in *Great Basin aquatic ecosystems history*, edited by R. Hershler, D. B. Madsen, and D. R. Currey. Smithsonian Contributions to Earth Sciences 33. Washington, D.C.: Smithsonian Institution Press.

Wolman, M. G. 1954. A method of sampling coarse river-bed material. *American Geophysical Union Transactions* 35:951–956.

Zdanowicz, C. M., G. A. Zielinski, and M. S. Germani. 1999. Mount Mazama eruption: Calendrical age and atmospheric impact assessed. *Geology* 27:621–624.

Chapter 4

Basin Sensitivity to Channel Incision in Response to Natural and Anthropogenic Disturbance

Dru Germanoski and Jerry R. Miller

Many upland streams in central Nevada are currently unstable and can be considered as systems in a state of disequilibrium. Whether the instability was initiated by natural disturbances such as climate change (Miller et al. 2001; chapter 2), lightning-generated wildfire (Germanoski and Miller 1995), heavy runoff associated with rainfall and snowmelt events (Germanoski et al. 2001) or anthropogenic influences, the dominant response of upland drainage basins to disturbance during the past 1,900 years has been channel incision. The most recent episode of incision in the watersheds began as early as 500 YBP (chapter 2). However, the nature of the responses between these basins is highly variable. Some basins are completely incised and appear to have reestablished channel morphologies that represent a new equilibrium. Others are (1) actively incising, (2) prone to catastrophic incision occurring over short periods (months to years), or (3) highly dynamic and experience significant changes in channel morphology during most major flood events.

The processes operating in streams of this area, and the channel morphologies produced by those processes, are the manifestations of the establishment of an equilibrium condition between channel morphology and the controlling factors. Although the exact nature of the equilibrium is dependent upon the time scale over which the channel is viewed (Schumm 1977), it is possible to recognize an equilibrium within the context of a specific time frame. For the purpose of this paper, channel equilibrium is defined as the maintenance of channel geometry where channel width depth ratio, elevation of the channel relative to the elevation of the valley floor, and channel planform geometry remains somewhat constant over time frames ranging from decades to centuries. This perception of an equilibrium form allows for dynamic change in the position of a channel through lateral migration or localized scour and fill associated with normal flood events. However, the vertical incision of a channel to depths several times the depth of the channel over a period of years to decades would be viewed as the result of disequilibrium between the controlling factors (e.g., a change in the rates of sediment or water delivery to the system) and channel form. The magnitude and type of channel change reflects the manner in which an entire basin (e.g., hillslopes) accommodate changes in external factors—a trait known as *basin sensitivity*. Variations in the nature and timing of channel response to disturbance suggest that drainage basins have varying degrees of sensitivity. Although the term *landscape sensitivity* can be interpreted in many ways (Allison and Thomas 1993), basin sensitivity is defined here as the tendency of a stream or river to respond to an environmental disturbance by going through a period of disequilibrium until a new equilibrium state is achieved. This definition is operationally similar to that presented by Schumm (1991), who defined geomorphic sensitivity as "the propensity of a system to respond to a minor external change." The concept of sensitivity arises from the recognition that many regions or landscapes are subjected to the same changes in external controls (tectonics, climate change, land use) but do not respond to the external stimulus at the same time or in the same manner (Schumm 1991). This suggests that different watersheds vary in their proximity to thresholds of instability (Patton and Schumm 1975). Ideally, the identification of a basin's sensitivity would make it possible to predict whether it is likely to respond to an external disturbance. Given that upland streams in the central Great Basin are prone to instability and that many streams are not fully incised (as evidenced by ongoing incision), placing the drainage basins into a sensitivity spectrum could provide land managers with a powerful tool for managing land use. Furthermore, the success of any attempt to restore an incised or actively incising channel could be assessed by a basin sensitivity rating and a projected trajectory toward an alternate equilibrium state. Streams that responded first, and have since established new equilibrium morphologies, may provide insight into the alternate states toward which actively incising or unincised streams are progressing.

Channel incision is problematic because it disrupts the physical and hydrologic systems (Miller et al. 2001; chapter 5), which in turn causes a change in the distribution and composition of vegetation within the riparian corridor, including that of the meadow complexes (chapter 7). The transformation of riparian wet meadow vegetation to dry land species assemblages represents a fundamental change in ecosystem structure and function (chapter 9).

During the past several years, we have studied a number of drainage basins in the Toiyabe, Toquima, Monitor, and White Pine Ranges to develop an understanding of the factors controlling the distribution of meadow complexes observed along the riparian corridor and the factors that influence channel incision and destabilization. The primary objectives of the research detailed in this chapter are to (1) evaluate unstable basins to determine the nature of channel response to natural and anthropogenic disturbance, (2) place unstable drainage basins into sensitivity categories that may be used for predictive purposes, and (3) develop an idea of alternate states toward which unstable channels are progressing.

Study Area

The Great Basin of central Nevada is characterized by north-south–trending fault-bounded mountain ranges separated by intermountain basins. A total of eighteen small (less than 100-square-kilometer) upland watersheds in the Toiyabe, Toquima, Monitor, and White Pine Ranges were studied to develop an understanding of drainage basin response to climate and land-use change (fig. 4.1). The drainage basins examined range from 4.9 to 87 square kilometers in area and from approximately 2,000 to 3,500 meters in elevation. Drainage basin relief varies between 300 and 1,500 meters. The mountain ranges are underlain by a combination of Mesozoic intrusive rocks, Tertiary volcanics (mostly welded tuffs), and early Paleozoic sedimentary and metasedimentary rocks (Stewart and McKee 1977; Kleinhampl and Ziony 1985).

The climate is arid to semiarid with average summer temperatures of approximately 20°C and average winter temperatures of –1°C. Average annual precipitation is approximately 40 centimeters, depending on elevation. Most of the precipitation falls as snow between November and May. June through October is relatively dry with occasional thunderstorms (Stewart and McKee 1977). Peak stream flow typically occurs in the spring in response to snowmelt, perhaps coupled with rainfall.

At low to middle elevations, the hillslopes are vegetated by Wyoming big sagebrush (*Artemisia tridentata wyomingensis*) interspersed with Utah juniper (*Juniperus osteosperma*) and single leaf pinyon (*Pinus monophylla*) woodlands. At higher elevations, mountain brush vegetation is dominant,

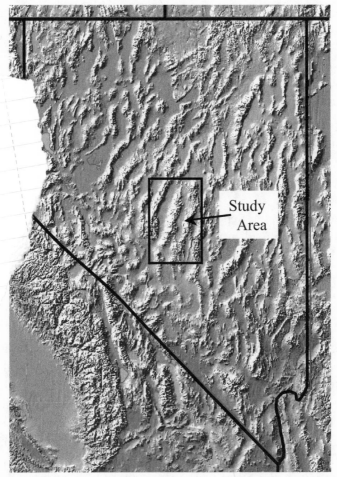

FIGURE 4.1. The location of drainage basins in central Nevada used in this study.

including mountain big sagebrush (*Artemisia tridentata vaseyana*) and snowberry (*Symphoricarpos oreophilus*). The riparian corridors range from narrow strips (less than 5 meters in width) to locally significant, broad meadow complexes characterized by shallow water tables, and dominated by species such as Nebraska sedge (*Carex nebrascensis*), Baltic rush (*Juncus balticus*), Kentucky bluegrass (*Poa pratensis*), creeping bent grass (*Agrostis stolonifera*), western blue flag (*Iris missouriensis*), and cinquefoil (*Potentilla gracilis*). Almost without exception, these meadow complexes are located upstream of side-valley alluvial fans that formed at the mouth of tributaries to the axial drainage (Miller et al. 2001; chapter 3).

The typical geomorphic architecture of the small upland watersheds consists of steep hillslopes with tors, extensive bedrock outcrops, and bedrock mantled by thin veneers of erosional pebble lag deposits with no significant pedogenic horizonation. Valley floors normally consist of moderately thick (greater than 6 meters) alluvial valley-fill deposits interfingered with side-valley alluvial fan materials. The upper several meters of fan and valley-fill materials were deposited during a period of drought and hillslope erosion dating from approximately 2500 YBP to 1900 YBP (Miller et al. 2001; chapter 3). The stratigraphic unit produced within the axial valley during this event is referred to as Qa_2, while the associated fan unit is named Qf_2.

In some cases, Qf_2 deposition extended across the valley and abutted against the opposite side-wall before being breached by the axial channels. These fan deposits significantly affect local topography of the axial valley and the dynamics of the axial channel, as is discussed in more detail below (fig. 4.2A). In other cases, the surface of side-valley fans grade to the alluvial fill without having a significant impact on the topography of the axial valley or the axial channel processes (fig. 4.2B).

The episode of hillslope erosion and valley filling between approximately 2500 and 1900 YBP followed a period of landscape and valley-floor stability that lasted from approximately 3500 to 2500 YBP (Miller et al. 2001; chapter 3). Landscape and valley floor stability prevailed during a period known as the Neoglacial, when cooler, wetter climatic conditions occurred in the Great Basin (Miller et al. 2001; chapter 3). This period of valley floor stability is marked by a well-defined paleosol (a buried soil) that developed in a valley-fill unit referred to as Qa_1. Qa_2 drought deposits overlie this paleosol. Where exposed by modern channel erosion in San Juan, Cottonwood, Stoneberger, and Barley Canyons, the Qa_1 paleosol is normally located close to the bankfull elevation of the modern channel.

The episode of upland erosion associated with the post-Neoglacial drought stripped the hillslopes of much of the fine-grained material leaving a pebble armor on the surface that is not mobilized by normal runoff events. Therefore, the predominant response to minor changes in climate associated with the Holocene or other disturbance since about 1900 YBP has been channel incision (Miller et al. 2001; Germanoski and Miller 1995). The most recent episodes of channel entrenchment may have been triggered by the Little Ice Age and a subsequent shift to more arid conditions (Miller et al. 2001).

A

B

FIGURE 4.2. (A) Side-valley fan prograding across the valley floor of Kingston Canyon; (B) side-valley fan graded to the valley fill in Kingston Canyon.

TABLE 4.1.

Summary of parameters measured or mapped along the study stream channels

Data Collected along Riparian Corridor	Watersheds with gaging stations	Other Basins
Knickpoint location and height	Illipah Creek	Barley Creek
Location and extent of overbank gravel deposits	Kingston Canyon	Big Creek
Number and distribution of stream terraces	Mosquito Creek	Birch Creek
Location of side-valley alluvial fans	Little Currant Creek	Cottonwood Creek
Location of "significant" bank erosion	Pine Creek	Indian Creek
	South Twin Creek	Johnson Creek
Data Collected at Selected Cross Sections	Marshall Canyon	
Bankfull width and depth	San Juan Creek	
Cross-sectional morphology	Stoneberger Creek	
Depth of entrenchment below valley floor	Upper Reese River	
Grain size distribution (optical method)	Veetch Creek	
Channel slope	Washington Creek	
Valley width		
Number and height of terraces above channel bed		

Methods

Data were collected from eighteen watersheds to describe their sensitivity to recent environmental alterations, expressed in terms of the rate of past changes in channel form and the potential for future adjustments in channel morphology. Data collection focused on the lower reaches of the riparian corridors and consisted of (1) delineating and mapping reaches with similar morphologic characteristics on aerial photographs and topographic maps, (2) mapping knickpoints, overbank deposits, and zones of bank erosion along the stream channels, and (3) measuring entrenchment depths, bankfull channel dimensions, channel slope, valley width, and bed material size for each of the delineated stream reaches. The mean size of the largest clasts at each cross section was determined by averaging measurements of the size of the largest ten clasts collected at each site. The median clast diameter was measured after the median clast size was selected by visual estimation of the clasts in the channel. Table 4.1 lists the parameters and the watersheds for which data have been collected.

The data demonstrated that the nature and magnitude of incision, the preservation of terraces, and the bed material size varied along the length of the channel. However, some short reaches exhibited significantly different characteristics than the channel as a whole. Thus, the mean of the net channel incision, grain size, and other parameters determined for a reach produced an unrealistic representation of the channel within the entire segment that was surveyed. To alleviate this problem, basin means for the measured parameters were weighted on the basis of reach length according to the following equation:

$$X_{i=1} = (\Sigma(RL_i)(x_i))/(TSL),$$

where X is the weighted average of the parameter of interest for the basin, RL is the length of a given reach, i, and is delineated in the field, x is the measured value for the parameter within the reach, i, and TSL is the total stream length that was mapped within the watershed.

The primary functions of a watershed are to move water and to move sediment to the basin mouth. The manner and rate in which this function is performed are revealed by the mean annual discharge, the shape of the flood hydrographs, basin lag times, and the amount of sediment eroded from basin slopes. One of the methods used to identify the geologic influences on basin functions is the analysis of basin morphometry. Basin morphometry is a collection of morphometric variables used to define linear, areal, and relief characteristics of a watershed. Increasingly, elements of basin morphometry are being related statistically to basin functions and used to predict or assess hydrologic and geomorphic processes, including flood potential and character (Patton and Baker 1976; Costa 1987; Patton 1988), sediment yields, and erosion rates (Gardiner 1990).

Basin morphometric data, including basin area, total stream length, drainage density, Shreve magnitude, and ruggedness numbers, were collected for the study basins using RiverTools (version 2.0) from U.S. Geological Survey (USGS) 7.5′ digital elevation models with a 30-meter resolution. Several additional parameters were obtained from 7.5′ topographic sheets, including basin shape, relief, and hypsometric integrals. Hypsometric analysis to obtain the hypsometric integrals provides insights into the distribution of mass in the watershed. It was hypothesized that hypsometric data would be useful in explaining both the pattern of incision within watersheds and the differences in incision observed between watersheds. Other morphometric parameters were calculated to evaluate the ability or potential for a stream to apply energy to the stream channel through the rapid delivery of snowmelt or precipitation. For example, total stream power is defined as the product of channel gradient and discharge (Bull 1979). Stream power per se could not be calculated because most of the drainage basins in this study are not gaged. However, because discharge should be proportional to basin area, basin area was utilized as a surrogate for discharge (Hack 1957; Lins-

TABLE 4.2.

Morphometric parameters measured for each drainage basin

Morphometric Parameter	Explanation or Equation of Definition
Basin Area (A)	Planimetric area of the drainage basin
Total Stream Length ($\sum L$)	Sum of the length of all channels in the drainage network within the basin
Drainage Density (D_d)	$\sum L/A$
Shreve Magnitude (N_1)	Number of first-order streams in the drainage basin
Relief (H)	Difference in elevation between the highest and lowest points in the drainage basin
Ruggedness R	HD_d
Relief Ratio (R_h)	H/L_b where L_b is the length of the drainage basin
Relative Stream Power ($\sum_r$)	AR_h
Basin Shape (S_b)	L_b^2/A
Hypsometric Integral (HI)	Percentage area under a dimensionless curve produced as the ratio of h/H and a/A where (a) represents the area of the basin above each elevation (h). High HI values indicate large portions of land are at high elevations in the drainage basin

ley et al. 1975). Similarly, relief ratio was used as a measure of average channel gradient (Strahler 1957). These substitutions allowed for the calculation of relative stream power for a given basin defined as the product of drainage area and relief ratio. The relative stream power estimated here should be proportionate to stream power and should serve as a reasonable substitute. Morphometric parameters used in this study are summarized in table 4.2.

In order to assess the controls on the degree of side-valley alluvial fan influence on axial channel dynamics, morphometric data (basin area, basin length, channel gradient, drainage density, relief, etc.) also were generated from 7.5′ topographic maps for forty-five side-valley fan basins that are tributary to Kingston Canyon.

The lithology of bedrock units that underlie the study basins was obtained from existing geologic maps of the area, including McKee (1976, 1:62,500), Dohrenwend et al. (1992, 1:62,500), Stewart and Carlson (1976, 1:250,000; 1978, 1:500,000) and Kleinhampl and Ziony (1985, 1:250,000). Wherever possible, the larger-scale maps were used to determine bedrock geology of the basins. In general, watershed divides were identified on the geologic maps, and the area of each lithologic unit within the basin of interest was determined with a digital planimeter.

Sensitivity Classification

All of the channels that were examined are prone to incision to varying degrees; however, there are distinct differences in the nature, timing, and rate of the response to disturbance. Therefore, drainage basins were divided on the basis of measurable parameters including the timing and degree of channel incision, the degree to which side-valley alluvial fans have influenced channel incision, and the presence of selected geomorphic features (e.g., knickpoints or overbank gravel deposits) that are indicative of channel instability. The use of these parameters to classify watersheds provides insights into basin sensitivity given the assumptions that (1) the most sensitive basins are those that respond first to disturbance, (2) channels that are the most responsive to a disturbance or high-magnitude runoff event are sensitive, and (3) channels that have maintained vestiges of their pre-disturbance morphology, and that are continuing to adjust, are less sensitive. On the basis of measured parameters, four groups of basins were delineated: Group 1, flood dominated; Group 2, deeply incised; Group 3, fan dominated; and Group 4, pseudostable.

Group 1, Flood-Dominated Channels

Group 1 drainage basins show evidence of significant localized incision, sediment mobilization, and redeposition that is commonly associated with channel avulsion and the destruction and transport of large woody debris during major floods (fig. 4.3). The net result is significant cutting and filling that creates multiple, discontinuous terraces produced during large flood events.

The intensity of coarse-grained sediment transport within these basins is illustrated by the USGS gaging station in Pine Creek. It is encased in a reinforced-concrete bunkerlike structure designed to withstand the force of high discharge and the impact of boulder bedload transport. The patterns of incision and deposition are chaotic along the riparian corridor, making it impossible to identify and delineate reaches with similar morphologic attributes for reach characterization. Therefore, this class of streams was not included in the quantitative analyses of channel response. Nonetheless, the qualitative assessments are consistent for all three basins that this category comprises: the Mosquito, Pine, and South Twin watersheds.

Valley bottoms within Group 1 basins have minimal sediment storage compared to many upland watersheds in the central Great Basin, and the valley floors consist of bedrock with a relatively thin to moderately thick veneer of cobble- to boulder-size alluvium that can be mobilized by flow events having an apparent return frequency of years to decades. These basins are underlain exclusively by Tertiary volcanic rocks (fig. 4.4) and are characterized by rugged, high-relief landscapes with moderate to high rela-

FIGURE 4.3. Boulder and woody debris deposits associated with a recent flood event in Pine Creek.

tive stream power (fig. 4.5). Morphologically, Group 1 basins have high hypsometric integrals, which indicates that a significant amount of landscape mass in the basins is at high elevations (fig. 4.5D). Also, these basins are not very elongated and have an average shape factor of 2.6 (fig. 4.5C).

The characteristics of Group 1 basins produce flood-dominated channels that are fundamentally unstable and susceptible to significant channel change attendant to nearly every major runoff event. The capacity of these channels to move large quantities of coarse bed material and the dynamic nature of these channels make them the most sensitive of all drainage basins in the area (fig. 4.3). Any land-use disturbance in Group 1 basins is likely to elicit an undesirable response and using bioengineering methods to stabilize the channels is unlikely to succeed.

Group 2, Deeply Incised Channels

Group 2 basins are characterized by deeply incised channels that have the greatest length-averaged incision values of all the streams in the data set (fig. 4.6A). For example, measurements from Barley Creek and Upper Reese River indicate that they are uniformly incised to depths ranging from 2 to more than 5 meters. Group 2 channels have eroded entirely through the side-valley alluvial fans. Thus, the influence of the fans on modern channel dynamics is minimal. Erosion through side-valley fans has

Drainage Basin Geology

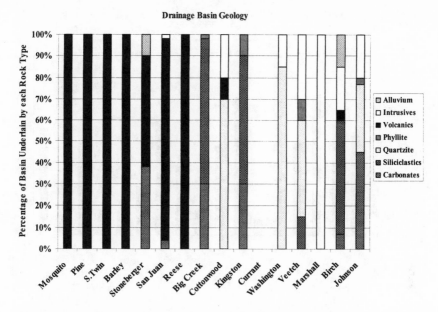

FIGURE 4.4. Histogram showing the relative percentage of lithotypes underlying each drainage basin.

smoothed the longitudinal profiles of the axial channels (fig. 4.6C) and exposed fan sediments up to several meters thick. Although the side-valley fans are not currently exerting a control on the main channels, several observations indicate that these fans extended across the valley and influenced the main channel in the past. The observations include (1) thick exposures (2 to more than 5 meters) of fan sediments in vertical cuts adjacent to the axial channels, and (2) projections of radial fan profiles that demonstrate that many side-valley fans extended across the main valleys, temporarily inhibiting the downstream flow of water and sediment. Channel bed incision and lateral bank erosion have produced vertical-walled channel trenches that are several meters deep and tens of meters wide. However, the active channels are typically 1 to 3 meters wide and normally have bankfull depths of approximately 0.5 meter deep (fig. 4.7). Although some of these channels possess knickpoints, which indicates that adjustments are still in progress, longitudinal profiles are relatively smooth, even through fan reaches. Also, the channels have uniformly spaced step-pool sequences in the upstream reaches and uniformly spaced pool-riffle sequences in the downstream reaches.

   Group 2 channels have the greatest number of terraces per reach length (fig. 4.6B). Terraces are produced either by discrete high-magnitude runoff

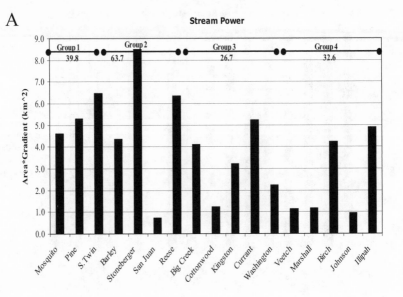

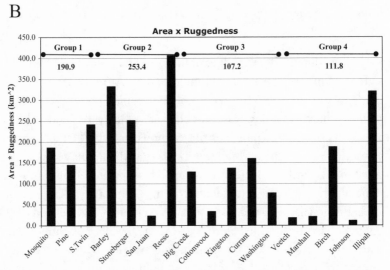

FIGURE 4.5. Composite plots of drainage basin morphometric characteristics: (A) relative stream power, (B) the product of area and ruggedness, (C) basin shape (length squared divided by area), and (D) drainage-basin hypsometry. Group averages are listed below the grouping bars.

events or by the passage of knickpoints associated with episodes of side-valley fan breaching. Although upland drainage basins are fundamentally unstable and prone to channel bed degradation, incision occurs episodically because the arid climate only produces significant runoff events on a decadal scale. Terraces in Group 2 systems are typically well preserved,

C

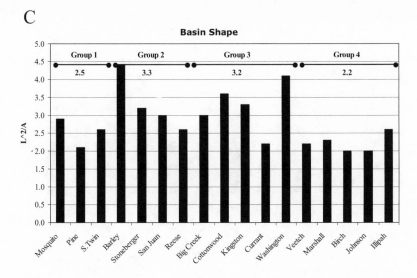

D

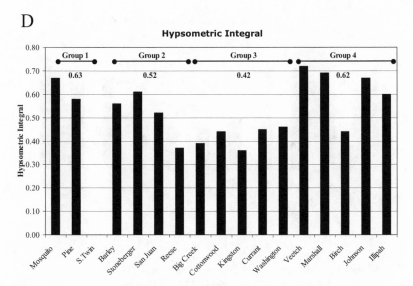

up to 30 meters wide at some locations, and contain abandoned channels (fig. 4.7). Terraces are particularly extensive and well preserved in the Upper Reese River, San Juan Creek, and Barley Creek Basins. Relatively wide valley floors and terraces inset within the channel trenches provide for the storage of large volumes of sand and gravel alluvium.

The deeply incised Group 2 channels are apparently the drainage basins that responded first and most significantly to the current episode of channel instability. This is evident by the smooth, longitudinal profiles of the incised channels and the development of mature vegetation in the inset

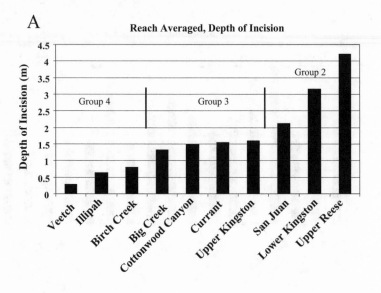

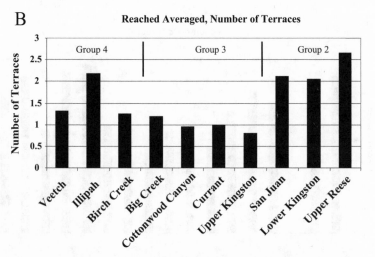

FIGURE 4.6. Composite plots of drainage basin response characteristics: (A) reach-averaged incision, (B) reach-averaged number of terraces, (C) degree of side-valley fan influence on channel gradient, and (D) reach-averaged degree of knickpoint influence.

riparian corridor. Group 2 drainage basins have many similarities with Group 1 channels in terms of geology and drainage basin morphometry. They both are underlain primarily by Tertiary volcanic rocks (fig. 4.4) and, with the exception of San Juan Creek, possess large, rugged, high-relief landscapes having moderately high hypsometric integrals and high relative stream power (fig. 4.5). These basins are among the most elongated in the data set with an average shape factor of 3.3 (fig. 4.5C).

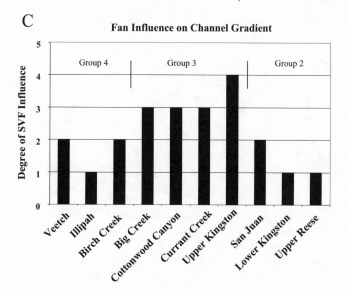

C
Fan Influence on Channel Gradient

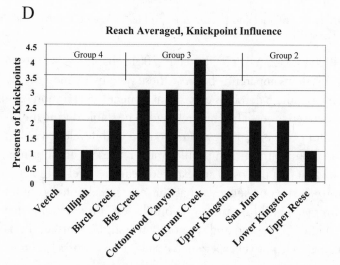

D
Reach Averaged, Knickpoint Influence

The fact that Group 2 channels have incised significantly in response to small-scale climate changes in the past indicates that these basins are highly sensitive to natural or anthropogenic disturbance. The development of step-pool and pool-riffle sequences and the establishment of vegetation on the low terraces and channel margins in these systems suggest that some of the streams in this category have established channel morphologies that are somewhat stable. These streams may represent longitudinal profiles and hydraulic geometries that are approaching a new, equilibrium state.

It is interesting to note that the buried soil produced during the period of landscape and valley-floor stability associated with the Neoglacial (Qa_1

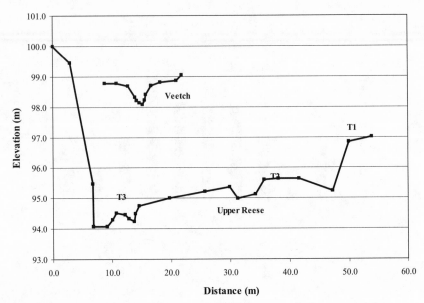

FIGURE 4.7. Channel and terrace cross profiles in Veetch Canyon and the Upper Reese River.

circa 4000 to 2500 YBP; Miller et al. 2001; chapter 3) is only exposed in Group 2 systems apparently because these systems are most deeply incised. In each case, the paleosol is exposed at or near the elevation of the bed of the modern channel. The channels had to be stable during the Neoglacial in order for pedogenic processes to produce well-developed soils (Miller et al. 2001). Thus, the stream profiles must represent equilibrium profiles. The fact that Qa_1 paleosols are exposed near the level of the channel bed in Group 2 streams suggests that the current channel profiles are approaching an equilibrium form. Nonetheless, these systems may experience additional incision and instability if disturbed further. In particular, San Juan Creek may yet experience instability as it works to eliminate steep reaches visible on its longitudinal profile.

The assertion that Group 2 (deeply incised) channels are approaching new equilibrium geometries raises the possibility that these systems should be subdivided into two sub-groups in a sensitivity classification scheme. Because these systems have responded first to the current episode of disturbance, this class of drainage basins should be considered sensitive to disturbance. Therefore, drainage basins in the region that have the same morphometric and geologic characteristics as Group 2 channels but have yet to incise should be considered very sensitive to disturbance because their geomorphic characteristics suggest that they are especially prone to

FIGURE 4.8. Step-pool sequence in the Upper Reese River.

incision. However, once completely incised, these systems may reestablish near equilibrium longitudinal profiles and approach new equilibrium channel morphologies. Therefore, once incised, these systems are much less susceptible to further disturbance. Drainage basins of this type may range in stability from unincised and very sensitive, to completely incised and approaching a new stable state. The assessment of a particular stream's relative position along the spectrum can be based upon whether (1) the stream's longitudinal profile has been smoothed or whether oversteepened reaches remain as sites of potential instability, (2) the low terraces have well-established stands of vegetation on their surfaces, and (3) the stream has a well-developed pool-riffle or step-pool sequence (fig. 4.8). Even though streams of this sort may have reached apparently stable alternative states, they may yet experience future instability associated with extreme runoff events; however, the magnitude of response should be minimal compared to the transformation these systems have already experienced.

Group 3, Fan-Dominated Channels

Group 3 fan-dominated channels are moderately incised with reach-averaged incision depths of approximately 1 meter (fig. 4.6A). Low incision values belie the fact that these channels are actively downcutting and, locally, these reaches are among the most dynamic in the study area. The dynamics and response of Group 3 basins are strongly influenced by side-valley alluvial fans. The most pervasive influence of the fans during the post-Neoglacial drought as well as in the present has been on the channel gradient (Miller et al. 2001; chapter 3). During the episode of fan building that occurred between approximately 2500 and 1900 YBP, the side-valley fans prograded onto the axial valley floors, reduced the valley gradient upstream, and blocked the transport of sediment down the trunk channels, causing aggradation of the valley floors upstream of prominent side-valley fans. Stratigraphic data constrained by radiocarbon dates demonstrate that fan sediments interfinger with valley fill sediments, and that widespread deposition was coeval on side-valley fans and along the axial channels (Miller et al. 2001). Pronounced sedimentation on side-valley fans during the post-Neoglacial drought has produced unstable longitudinal profiles that are relatively flat upstream of the fans and steepen significantly where the channels traverse the fans (fig. 4.9). The oversteepened segments on the downstream sides of the fans are inherently unstable and prone to channel incision because the steeper slopes produce high shear stresses on the channel bed. Although the steep gradients produce high shear stresses that promote downcutting, channel incision through fan sediments has been limited by relatively coarse alluvium eroded from the fan deposits (fig. 4.9). However, once an episode of incision occurs within a fan segment, the wave of incision migrates up channel into the lower gradient reach of the valley fill, producing erosional inset terraces. The depth of entrenchment accompanying each episode of channel incision is controlled largely by the development of boulder-defended knickpoints. These knickpoints serve as local base levels that limit the depth of incision in the channel segment immediately upstream. Therefore, the elimination of fan control on upstream reaches usually occurs through multiple episodes of knickpoint migration and channel bed incision rather than as a single event.

A knickpoint in Kingston Canyon migrated several hundred meters over a period of approximately four years (1997–2001). There is no evidence of the knickpoint at the present time; however, its passage is marked by a channel that has been incised as much as 1 meter throughout most of the reach and by an inset terrace that represents the former channel bed. As of 2002, there is a pronounced boulder-armored knickpoint with over 1 meter of relief positioned within the fan segment below this recently incised reach of stream in Kingston Canyon. Once this armor is broken, another distinct phase of channel bed degradation and bank collapse will occur as this knickpoint migrates upstream. In some segments, two terraces may be

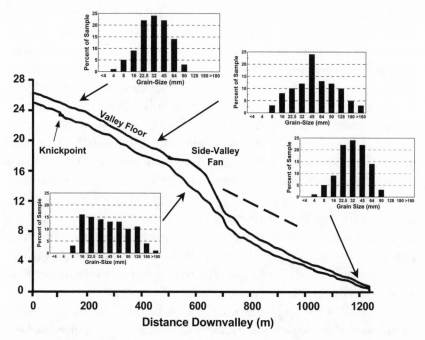

FIGURE 4.9. Longitudinal profile of the axial channel of Kingston Creek where it crosses a prominent side-valley fan. Inset histograms show the grain-size distribution of sediment on the channel bed upstream of the fan, at the fan, and downstream of the fan.

preserved that mark two episodes of channel cutting, but in some instances the terrace has been destroyed by bank collapse and channel widening.

Kingston Canyon and Big Creek are among the best examples of fan-dominated basins. Both of these watersheds possess meadow complexes that are still intact upstream of prominent side-valley fans, although the vegetation composition along the riparian corridors is changing as the channels incise and the groundwater tables drop (chapter 7). Reach-averaged incision values average approximately 1.5 meters (fig. 4.6A). The relatively shallow depths of incision primarily result from the difficulties involved in transporting coarse fan gravels and, in some cases, exhumed bedrock (fig. 4.6C, 4.6D). Locally, however, the fan deposits have been eroded, allowing channel bed incision to occur through the upstream migration of knickpoints. In cases where meadow complexes are present immediately upstream of a side-valley fan, channel bed degradation is aided by the release of groundwater from the subsurface reservoir. Groundwater seepage and erosional sapping is, in fact, a primary mechanism of bank erosion, channel widening, and terrace destruction in these environments.

Although fan-dominated drainage basins are inherently unstable and prone to incision and widening, the fan gravels effectively protract the basin response times to disturbance. In essence, the systems are metastable, meaning that they experience multiple episodes of incision that slowly create entrenched channels as the systems work to smooth their longitudinal profiles.

The geology of Group 3 basins is significantly different from Groups 1 and 2. Groups 1 and 2 are underlain by Tertiary volcanic rocks, whereas Group 3 drainage basins are underlain primarily by sedimentary and metasedimentary rocks (fig. 4.4). On average, fan-dominated basins are less rugged, have lower relief, and possess lower relative stream power than Group 1 and 2 channels (fig. 4.5). Fan-dominated drainage basins are elongated and have an average drainage basin shape value of 3.2. The shape factor (3.2) is indistinguishable from the value of 3.3 measured for Group 2 channels (fig. 4.5C). The Group 3 drainage basins also have low hypsometric integrals, which means that significant mass has been removed from these basins by erosion.

As a result of the stabilizing influence of the side-valley fans, Group 3 basins are the least sensitive in the region. However, it is important to keep in mind that although these channels are experiencing the slowest rates of incision and have incised the least, they are fundamentally unstable and will continue to incise whenever dissection of the side-valley fans can occur.

Comparison of Group 2 and Group 3 basins indicates that they once had similar valley floor morphologies and that the side-valley fans in Group 2 basins had a significant influence on channel processes. This leads to the conclusion that Group 3 channels will eventually be transformed into incised streams with relatively smooth (and stable) longitudinal profiles (figs. 4.7 and 4.9).

Group 4, Pseudostable Channels

Although Group 4 basins have experienced minimal incision, they have the potential to rapidly and catastrophically incise. Thus, they are designated as pseudostable. Reach-averaged incision rates in Veetch, Illipah, and Birch Creeks are typically less than 1 meter (fig. 4.6A). Channel bed erosion in these systems is localized rather than continuous with incised reaches separated from unincised reaches by prominent knickpoints whose upstream migration is aided significantly by groundwater sapping. Within the incised segments, these channels have discontinuous terraces produced by the migration of knickpoints (fig. 4.6B, 4.6D). Unincised reaches are characterized either by well-defined, somewhat permanent channels set 1 meter or so into the valley floor (fig. 4.7) or by indistinct, shallow channels flowing across flat valley floors or meadows. Birch Creek is typical of this category of drainage basin. The lower portion of the meadow in upper Birch Creek is incised to a reach-averaged depth of approximately 1 meter. The incised segment is either (1) flanked by a well-defined terrace that includes local, well-preserved cutoff meander scars or (2) characterized by a narrow channel with vertical banks and no terraces. This incised reach is separated from a shallow, stable, grass-lined channel by a pronounced knickpoint with 1.5 meters of relief. Knickpoint migration is apparently aided by groundwater sapping by (1) water tables that are at the surface throughout most of the meadow upstream of the knickpoint, (2) knickpoint faces that have a concave geometry, and (3) seepage of water along the face of the knickpoint.

Other systems that fit in this category have incised very rapidly in response to short-term forcings such as wildfire or exceptional runoff associated with snowmelt and rainfall. For example, Crow and Marshall Creeks were locally eroded to depths as great as 4 meters in response to a lightning-generated wildfire that occurred in 1981 and that was followed by two years of above-average precipitation (Germanoski and Miller 1995). Aerial photo coverage of Crow Canyon revealed that it was drained by a shallow, stable channel flowing across a nearly flat, meadowlike valley floor prior to the erosional events of 1982 and 1983. By the end of 1983, the axial channel was completely entrenched to an average depth of 1.9 meters (Germanoski and Miller 1995). Marshall Creek also was incised in 1982–1983, but significant channel bed erosion was restricted to two reaches rather than extending throughout the axial valley.

Marshall Creek was more deeply eroded in 1998 over a period of days to depths of up to 6 meters as a result of heavy rainfall and snowmelt. Erosion occurred in large part by groundwater sapping and knickpoint migration (Germanoski et al. 2001) that was concentrated in two discrete segments of the axial channel. The sediment produced by incision was deposited as channel fill or upon the valley floor downstream of the erosional reaches (Germanoski et al. 2001). The extremely rapid incision of Crow and Marshall Canyons illustrates that drainage basins in category 4 are temporarily stable and prone to almost instantaneous reaction to natural or anthropogenic disturbance.

Group 4 drainage basins are underlain primarily by intrusive igneous and sedimentary rocks (fig. 4.4). These lithotypes weather to pebble- and fine-gravel-sized material that produces valley fills and channel bed sediment that are relatively fine grained and highly erodible. With the exception of Birch Creek and Illipah Creek, Group 4 basins tend to be less rugged, have lower relief, and have lower relative stream power than other drainage basins in the study (fig. 4.5). Birch Creek and Illipah Creek have higher stream power and area-relief values primarily because they are large basins. Group 4 basins exhibit more-equidimensional morphologies in plan view and have an average drainage basin shape value of 2.2 (fig. 4.5C). These basins also have high hypsometric integrals similar to Group 1 basins (fig. 4.5D).

Group 4, pseudostable channels are the most enigmatic in the study area because many streams in this category are either unincised, or slightly incised, while others have incised rapidly in response to apparent threshold-crossing disturbances or runoff events. Because these basins can be eroded to depths of several meters below the valley floor in a matter of days or months, they must be viewed as being very sensitive to human or climatic disturbance. For example, basins such as Veetch Creek are apparently stable and only slightly incised. However, Crow Creek, a basin with similar characteristics to Veetch Creek, was also unincised and apparently stable prior to being burned by wildfire in 1981. Following the wildfire and two years of above-average precipitation in 1982 and 1983, Crow Creek incised as much as 3.9 meters (Germanoski and Miller 1995). Similarly, several segments of Marshall Canyon, which was also burned in 1981, were incised significantly in the years following the 1981 wildfire. Other reaches of this channel remained stable until 1998 when the channel was eroded tremendously through processes of groundwater sapping and high runoff associated with a large snowpack and a wet spring (Germanoski et al. 2001). Thus, Group 4 drainage basins have the capacity to respond quickly and significantly to disturbance.

Table 4.3 provides a summary of the geomorphic characteristics of each of the basin sensitivity groups.

Factors That Control Drainage Basin Sensitivity to Disturbance

The previous section demonstrated that upland watersheds in central Nevada can be categorized according to their sensitivity to respond to both natural and anthropogenic disturbance. The classification system relied on a number of parameters that require considerable time and effort to document. Moreover, field work is required for data collection, and it could be argued that the information should be collected by geomorphologists

TABLE 4.3.

Drainage basin group characteristics

Category	Geology	Relief Characteristics	Basin Shape	Fan Influence and Longitudinal Profile	Incision and Terraces
Group 1 Flood dominated	Volcanic rocks	Rugged High stream power High hypsometric integrals	Equant Low shape factor	Low fan influence	Localized incision Discontinuous terraces
Group 2 Deeply incised	Volcanic rocks	Rugged High stream power High hypsometric integrals	Elongated High shape factor	Low fan influence Smoothed longitudinal profile	Deeply incised Many continuous terraces
Group 3 Fan dominated	Sedimentary and metasedimentary rocks	Low stream power Low hypsometric integrals	Elongated High shape factor	High fan influence Stepped longitudinal profile	Localized incision Discontinuous terraces
Group 4 Pseudostable	Intrusive igneous and siliciclastic rocks	Moderate to high stream power High hypsometric integrals	Equant Low shape factor	Low fan influence Stepped longitudinal profile	Localized deep incision Discontinuous terraces

or other individuals with experience in analyzing fluvial processes and alluvial stratigraphy. Thus, for the classification scheme to be useful as a tool
for land management or restoration, the controls on sensitivity need to be
determined, particularly those associated with basin morphology and geology that will allow the watersheds to be classified using existing maps and
other data sets. Identifying the controls on basin sensitivity has the added
advantage of providing for a more detailed understanding of the factors that
are directly affecting geomorphic responses to environmental change. The
following discussion examines the controls on basin sensitivity at a watershed scale.

Controls on the Degree of Side-Valley Fan Influence

Side-valley alluvial fans currently play a fundamental role in determining
the morphology of the valley floor and the local dynamics of channel systems in Group 3 drainage basins, and they clearly played a similar role in
Group 2 basins in the recent past. Conversely, side-valley fans have only a
limited impact on channel dynamics within Group 1 and Group 4 basins.
Therefore, in order to fully understand the differences in basin sensitivity,
and the variations in response to disturbance among the various drainage
basin types, it is essential to identify the factors that determine whether side-
valley fans are prominent features within any given watershed. Qualitative
observations suggested that large, elongated tributary valleys with relatively
gentle gradients and large sediment-storage capacities tend to produce fans
that grade into the valley floor of the master channel, whereas smaller,
more equant (valley width is more similar to valley length) tributary valleys
having steep gradients tended to produce fans that prograde onto the valley floor and influence the geomorphic processes in the master channel.

Kingston Canyon provides a unique setting in which to gain insight into
the controls on side-valley fan development and morphology because the
forty-five major fans in the watershed form a continuum between two distinct morphological types. The end members of the continuum consist of
(1) fans that form prominent landforms that prograde into the axial valley
and exert a significant influence on the master channel, and (2) fans that
grade smoothly onto the alluvial valley floor without significantly altering
the local valley gradients, or impacting axial channel processes (fig. 4.2).

In order to evaluate the potential morphometric control on side-valley
fan morphology, the forty-five fans in Kingston Canyon were classified in
the field into two end-member and two additional transitional categories
including (1) prograding fan, (2) graded fan, (3) prograding-graded fan, and

TABLE 4.4.

Morphometric characteristics of Kingston Canyon tributaries that have alluvial fans

Fan Type	Area (km^2)	Relief (m)	Length (km)	Gradient
Graded fans	2.72	735	2.91	0.29
Prograding fans	0.67	546	1.24	0.49
Transitional fans/prograding	1.84	594	2.18	0.32
Transitional fans/graded	1.76	673	2.01	0.36

(4) graded-prograding fan. The first term in the latter two transitional categories describes the predominant characteristic of the fan. Quantitative evaluation of the morphometric characteristics of the tributary basins confirms the qualitative observations that short, steep, tributary basins produce fans that prograde into the axial valley and that elongated, low-gradient tributaries produce fans that grade to the surface of the valley fill without influencing the master channel. Prograding fans have drainage basins with an average length of 1.24 kilometers, whereas graded fans have an average length of 2.91 kilometers (table 4.4). Likewise, prograding fans have drainage basins with an average gradient of 0.49 kilometer per kilometer, whereas graded fans have an average gradient of 0.29 kilometer per kilometer (table 4.4). Fans having transitional morphologies fall in between these extremes and are morphometrically indistinguishable from one another (table 4.4). Prograding fans also exhibit higher relief and tend to be smaller in size (table 4.4).

Further consideration of these relationships suggests that for any given drainage basin the fundamental control on side-valley drainage basin morphology is basin width. In an area having semi-uniform relief, narrow drainage basins will have relatively short, steep, tributary basins with high gradients. Thus, the elongated drainage basins should have more numerous short, steep, tributary basins that will produce a high percentage of side-valley alluvial fans that prograde onto the valley floor and interact with the axial channels.

Drainage-basin shape was calculated as the dimensionless ratio L^2/A. The higher the shape factor, the more elongate the drainage basin. Figure 4.5C shows that Group 2, fully incised basins, and Group 3, fan-dominated drainage basins, are the most elongate watersheds with average shape factors of 3.3 and 3.2, respectively. Channels within both of these categories of drainage basins are experiencing (Group 3), or have previously experienced (Group 2), significant impact from side-valley alluvial fans on their behavior. Groups 1 and 4 show no evidence of side-valley influence because these basins are relatively wide and have basin shape values of 2.6 and 2.2, respectively (table 4.2). Sediment produced by side-valley tributary channels in Group 1 and 4 basins is deposited either in the tributary basin itself or on an alluvial fan that is located close to the valley margin where it is too far removed from the axis of the main valley to affect the axial channel. Therefore, the primary control on the degree of side-valley fan influence on axial channel behavior appears to be drainage basin shape.

Morphometric and Geological Controls on Channel Response to Disturbance

Although the effects of basin shape on side-valley fan morphology can be clearly defined, other relationships between drainage basin morphology, geology, and sensitivity are less obvious. Nevertheless, careful evaluation of the morphometric and geological characteristics of the drainage basins reveals that while relationships are complex, they can be delineated. Much of the complexity results from interactions between variables that operate synergistically to facilitate channel instability. For example, a number of factors can interact to promote incision, and the entrenchment observed in different basins, or even different reaches of the same basin, can be produced by different combinations of the variables.

Any analysis of drainage basin sensitivity must consider factors that establish a balance between forces that cause and forces that resist erosion of the channel bed and banks. The forces of erosion can be expressed in terms of stream power (Bull 1979), or as relative stream power as used in this study. Although discharge should be proportional to drainage basin size, other aspects of basin morphology will determine peak discharge and, thus, stream power. For example, variables that commonly control the rate at which runoff reaches the master channel include geology, area, basin shape, relief, drainage density, ruggedness, and channel slope. In this investigation, several morphometric variables have been combined to create parameters that summarize the most significant variables influencing the magnitude of the peak flows. High relief and drainage density are likely, for instance, to produce relatively high peak flows, but the total discharge associated with a snowmelt or rainfall event would also be determined in large part by basin size. Therefore, these three terms were combined as the product of area, relief, and drainage density (area-ruggedness). Similar parameters have been produced in the past to describe the likelihood that a high peak discharge will be produced during a runoff event. An example is the ruggedness number, which is defined as the product of drainage density and relief and which is thought to be a measure of average slope within the basin (Strahler 1957).

The morphometric data also reveal that the most sensitive basins (Groups 1, 2, and 4) are characterized by high hypsometric integrals (fig. 4.5D). High hypsometric integrals reflect drainage basins that have large portions of landscape at high elevation. Large areas of land at high elevation can capture and retain snow, and produce high snowmelt discharge in the spring and early summer. Therefore, high hypsometric integrals (integrals greater than 0.5) can be used to identify basins that are likely to be sensitive to disturbance.

The lithology of the bedrock underlying the watershed affects basin sensitivity in two significant ways: (1) bedrock geology determines the infiltration capacity/runoff ratio by influencing the hydraulic conductivity of the surface and subsurface materials, and (2) geology influences the grain size and erodibility of debris mantling the hillslopes and the channel bed. Although few of the drainage basins in the study are gaged, the limited data that are available suggest that basins underlain predominantly by volcanic rocks produce runoff events that lead to higher stream powers and more dynamic channels. Compare, for instance, the hydrographs for a basin underlain primarily by carbonate and siliciclastic rocks to a basin underlain predominantly by volcanic rocks (fig. 4.10). The basin underlain by volcanic lithologies is characterized by a shorter lag time, higher peak flows, and shorter-duration high flows than those associated with the other rock types (fig. 4.10).

The relationships among stream power, area-ruggedness products, hypsometric integrals, and drainage basin sensitivity are complicated, but, on average, the highly sensitive streams that comprise Groups 1 and 2 basins are characterized by relatively high stream power, high area-ruggedness products, and high hypsometric integrals (figs. 4.5A, 4.5C, 4.5D). Thus, these are relatively large, steep basins that have the capacity to transfer water from the landscape to the master channels quickly. The exception to this rule is San Juan Creek. Part of the explanation for San Juan Creek having been thoroughly incised is the relatively small grain size of the channel bed material (D50 averaging 5.2 centimeters), which is derived from the welded-tuff that underlies most of the basin. Therefore, although the basin morphology and size does not particularly promote high peak flows, the channel bed is unusually erodible because of relatively fine-grained bed material. Group 1 and 2 systems also are prone to incision because

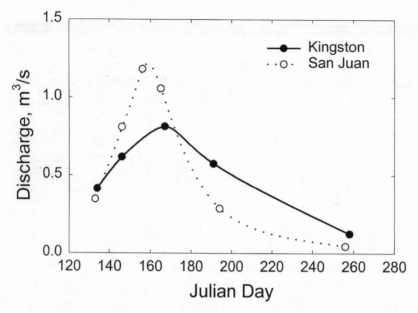

FIGURE 4.10. The hydrograph of a drainage basin underlain by volcanic rocks (San Juan Creek) compared with the hydrograph of a basin underlain by sedimentary and metasedimentary rocks (Kingston Canyon).

these basins are underlain almost exclusively by volcanic rocks, which produce runoff and high relative stream power (figs. 4.5A, 4.10). With regard to geology, Groups 1 and 2 are readily distinguished from the rest of the drainage basins in the study.

There also are other factors that may explain why San Juan Creek basin, characterized by a relatively small basin area, low area-ruggedness products, and low relative stream power is so intensively incised. Recent work by Thomas Bullard and Daniel Lahde (personal communication) suggests that incision in that basin was influenced by the failure of beaver dams in some areas and by road capture of incising channels in other segments of the drainage network. Observations of the erosive impact of beaver dam failure have also been made in Cottonwood Creek and of road capture in Stoneberger Canyon. These observations indicate that other factors may exacerbate channel instability on a site-specific basis in these inherently unstable drainage basins.

Group 3 basins that are significantly influenced by side-valley fans typically have lower relative stream power, lower area-ruggedness products, and lower hypsometric integrals than the more fully incised channels in Group 2 (figs. 4.5A, 4.5C, 4.5D). However, Group 2 and Group 3 drainages also differ geologically. Thus, the inability to incise through the side-valley fan deposits in Group 3 basins apparently results from the combined effects of basin morphology and basin lithology. These basins are most stable because the effects of side-valley alluvial fans forestall channel incision and protract the transformation of the channel in response to disturbance.

Some Group 4, pseudostable channels, have limited stream power as the result of low gradients and smaller basin size, whereas others, notably Birch and Illipah Creeks, have high relative stream power that is primarily generated because of relatively large basin areas. The explanation for the unique behavior of these systems seems to be a combination of basin shape and basin geology. These basins are all equal dimensional in shape (fig. 4.5C), a morphometric property that facilitates high peak flows. In addition, significant portions of these basins are underlain by intrusive igneous rocks, chiefly granites and quartz monzonites (Stewart and McKee 1977; Kleinhampl and Ziony 1985). These rock types weather to produce granule-sized material, *grus*, that mantles the hillslopes and comprises the valley fills. Grus facilitates infiltration and groundwater seepage through the valley fill, both of which favor groundwater flow and erosion through groundwater sapping. Further, the noncohesive nature of grus in the valley fill may facilitate rapid erosion by sapping in response to rapid melting of a large snowpack or sustained high rainfall (Germanoski et al. 2001).

Knickpoints can be maintained in these basins for significant time periods by surface layers of cohesive, fine-grained sediments armored by dense root mats consisting of meadow grasses. Nonetheless, observations and analysis of sequential aerial photographs demonstrate that channels in these basins can be incised very rapidly (weeks to years) by knickpoint migration associated with groundwater sapping processes.

Management and Restoration Implications

Small upland watersheds in the central Great Basin are inherently unstable and have been prone to incision for at least the past four hundred to five hundred years. However, the timing and rate at which channels incise depend on their sensitivity to disturbance, which in turn, is controlled by the geology and morphology of the watershed and the degree to which side-valley alluvial fans influence the axial channel dynamics. Once initiated, channel incision can be exacerbated locally by road capture, beaver dam failure or other natural or anthropogenic influences. In the arid Great

Basin, wildfire can be the final catalyst for destabilization and channel bed erosion (Germanoski and Miller 1995).

Drainage basins can be separated into four sensitivity classes on the basis of criteria that describe the magnitude, timing, and rate of channel incision along the riparian corridor. The most sensitive channels include those that experience major channel bed change (incision, deposition, channel avulsion, and realignment) in response to every low-frequency, high-magnitude runoff event (recurrence intervals on the order of decades) (Group 1 channels); those that respond catastrophically (meters of incision) to threshold-crossing runoff events (Group 4 channels); and those that have responded to recent climate change or other disturbances by transforming completely through deep incision and the establishment of an alternate stable state morphology (Group 2 channels). The most stable channels are those that are significantly influenced by side-valley alluvial fans (Group 3 channels) because coarse-grained fan deposits act as local base-level controls that help maintain channel gradients upstream of the fan and resist incision. It is important to recognize that the stabilizing effect of the fans is a temporary impediment to channel incision rather than a truly stabilizing influence.

Group 2 channels provide insights into the alternate morphology toward which the incising channels are progressing, and were among the first basins to respond to the current episode of disturbance. Thus, they provide insight into the controls on basin sensitivity. With respect to the fate of the fan-dominated basins, there is abundant evidence that side-valley fans once played a role in controlling channel dynamics in Group 2 (deeply incised basins) as they are currently doing so in Group 3 basins. However, the fact that the side-valley fans have been breached and the axial profiles smoothed by erosion demonstrates that ultimately, side-valley fan influence will be minimized or eliminated in the fan-dominated basins. Group 2 channels are inset several meters into the valley fill, flanked by one or more pronounced terraces, and have relatively smooth longitudinal profiles with step-pool or pool-riffle sequences. The lowered water tables in these systems support phreatophytes and other water-loving vegetation in narrow strips within the alluvial trench. In contrast, Group 3 systems often support broad meadows that extend across much of the alluvial valley floors immediately upstream of side-valley alluvial fans (chapter 7). The exposure of the Qa_1 paleosol at or near the elevation of the incised channels may be a coincidence, but it is more likely an indication that once the unstable channels have re-incised to the level of the Neoglacial-era channel beds, they have reestablished sufficiently smooth longitudinal profiles to maintain some measure of channel bed stability. Once this new configuration is established, it is unlikely that these systems will experience additional significant bed erosion unless they are affected by another major change in the environment.

Parameters such as the product of area and ruggedness and relative stream power (area × relief ratio) are measures of a basin's ability to concentrate runoff and energy on the valley floor during a snowmelt or precipitation event and represent primary controls on channel sensitivity. The most sensitive systems are those characterized by large, steep, rugged basins underlain by volcanic rocks. The importance of the combined effect of geology and morphology is best illustrated by comparing Group 2 and Group 3 systems. Group 2 and 3 channels have similar elongated shapes, which facilitates side-valley fan progradation and influence on axial channel dynamics. However, because of the effects of volcanic geology on hydrology, and the effects of basin morphology and hypsometry on stream power, Group 2 channels rapidly incised through the side-valley fans.

Geology plays a prominent role in determining the sensitivity of Group 4 basins. The common attribute of these basins is the presence of intrusive igneous rocks in the watersheds that produce a granular valley fill consisting in part of noncohesive pebbles (grus) that facilitate infiltration and erosion by groundwater sapping. It is interesting to note that the only stream in the study that is underlain by a significant area of intrusive igneous rocks that does not fit in category 4 is Cottonwood Creek. Despite the geologic similarity to Group 4 systems, the elongate shape of the basin results in a significant influence by side-valley fans.

The results of this study indicate that it is possible to produce a sensitivity hierarchy for inherently unstable upland drainage basins in the central Great Basin. However, the controls on basin sensitivity are complex. A "cookbook" approach to stream classification and sensitivity analysis will not work in these basins. Sensitivity assessment requires geomorphic field experience, a more than cursory understanding of geomorphic principles, and a significant knowledge of the nuances of the drainage basins in this region. This increases the challenge for both land managers and scientists working in this region.

These results are consonant with other attempts to categorize landscapes within a sensitivity framework. Other research efforts on landscape sensitivity also have demonstrated that the inherent complexity and multivariate nature of geomorphic systems requires that (1) sensitivity analyses must be based on multiple indicators (Downs and Gregory 1993) and (2) sensitivity predictions are complicated by the basin's geomorphic history, land use, climate change, scale, and other factors (Schumm 1985, 1991; Blumler 1993; Boardman 1993; Brunsden 1993; Burt et al. 1993; Derbyshire et al. 1993; Evans 1993; Gerrard 1993; Perkins and Thomas 1993; Roberts and Barker 1993; Shaw and Thomas 1993). Although most efforts to assess landscape sensitivity are successful at explaining variations in response to disturbance (Gerrard 1993; Roberts and Barker 1993), few have been able to quantify the degree of sensitivity of a basin prior to instability. The degree to which this is possible appears to be inversely proportional to the number of variables that must be taken into consideration in the analyses. For example, Patton and Schumm (1975) were able to clearly identify thresholds of landscape stability with respect to gully development in an area of uniform geology, climate, vegetation, and land use. However, as the number of variables increase, the more complex the relationships become and the more difficult it is to use the classification scheme for predictive purposes (Schumm 1985, 1991; Derbyshire et al. 1993; Perkins and Thomas 1993). Although it is not possible to precisely quantify the degree of basin sensitivity to disturbance in the Great Basin, sensitivity analyses based upon past response to disturbance, and the controls on channel response, can be a useful first step to making management decisions and assessing the likelihood of future change needed for restoration designs (Schumm 1991; Blumler 1993; Boardman 1993; Brunsden 1993; Burt et al. 1993; Derbyshire et al. 1993; Downs and Gregory 1993; Evans 1993; Gerrard 1993; Perkins and Thomas 1993; Roberts and Barker 1993; Shaw and Thomas 1993).

Despite these complexities, it is clear that drainage basins that have morphological and geological characteristics that operate together to produce rapid runoff are most sensitive to disturbance. The critical attributes are high hypsometric integrals (greater than 0.5), high area-ruggedness products, high relative stream power (greater than approximately 30 square kilometers), and volcanic rocks. In contrast, basins that are highly elongated (L^2/A values greater than 3) and possess low hypsometric integrals (less than 0.5) tend to be the most stable and offer the greatest chance for restoration.

It is critical to recognize that sensitivity assessments for land management purposes in the central Great Basin should proceed by intensively examining each watershed rather than by trying to place a drainage basin into one of the four identified categories. Drainage basins are unique and, therefore, must be evaluated individually. Nonetheless, the data presented above provide for an understanding of the relative importance of selected parameters to basin sensitivity and provide a template for the types (and order) of analyses that should be carried out. As an example, the above data indicate that bedrock geology has a significant influence on basin sensitivity and, therefore, the first parameter to consider is basin lithology. Drainage basins that are underlain by volcanic rocks or coarse-grained intrusive rocks should be delineated as potentially sensitive to disturbance. The next set of parameters to consider could include hypsometry, area-ruggedness, and stream power. If the measured values are high, then the potential for the basin to be sensitive increases because high values for these parameters indicate that the stream has high erosive capability. Next, one could measure the drainage basin shape to determine if there is a high likelihood that side-valley alluvial fans are significant factors inhibiting or protracting channel erosion. Assuming that the drainage basin is elongated and has a shape factor greater than 3, which suggests that side-valley fans play a significant role, the system would have to be evaluated in the field to determine whether side-valley fans prograde across the valley and forestall channel erosion or have been breached by channel incision. If the system has a high potential for erosion and the fans have not been breached, the systems would be considered very sensitive to disturbance (these would be systems having all of the characteristics of Group 2 basins, except that the fans have yet to be eroded). If on the other hand the fans have been eroded and the channel has downcut several meters below the valley floor, thereby creating a smooth longitudinal profile with well-developed step-pool sequences, one might conclude that the system, although initially sensitive to disturbance, is now in an alternate state morphology that is somewhat stable and insensitive to further disturbance. The point is that the sensitivity analysis described in this chapter is useful only when the entire spectrum of geologic and geomorphic parameters that influence basin sensitivity are examined one basin at a time.

Acknowledgments

Many individuals aided in the collection of field data for this investigation. They include Lafayette College students Nathan Hawk, Brian Kortz, Daniel Latham, and Carolyn Ryder, and Western Carolina University students Kurtis Beshers, Chris Bocchichio, Christiana Bruinsma, Ronnie Dilbeck, Chris Ernst, Sara Vartabedian, and Rob Vartabedian. We greatly appreciate their help, without which we would have been unable to complete the investigation. This chapter was greatly improved by comments made by reviewers Kyle House and Dale (Dusty) Ritter. We would also like to thank the USDA Forest Service for providing both financial and logistical support.

Literature Cited

Alison, R. J., and D. S. G. Thomas. 1993. The sensitivity of landscapes. *Landscape sensitivity.* West Sussex, UK: John Wiley & Sons.

Blumler, M. A. 1993. Successional pattern and landscape sensitivity in the Mediterranean and Near East. Pp. 287–305 in *Landscape sensitivity,* edited by R. J. Allison and D. S. G. Thomas. West Sussex, UK: John Wiley & Sons.

Boardman, J. 1993. The sensitivity of downland arable land to erosion by water. Pp. 211–228 in *Landscape sensitivity,* edited by R. J. Allison and D. S. G. Thomas. West Sussex, UK: John Wiley & Sons.

Brunsden, D. 1993. Barriers to geomorphological change. Pp. 7–12 in *Landscape sensitivity,* edited by R. J. Allison and D. S. G. Thomas. West Sussex, UK: John Wiley & Sons.

Bull, W. B. 1979. Threshold of critical power in streams. *Geological Society of America Bulletin* 90:453–464

Burt, T. P., A. L. Heathwaite, and S. T. Trudgill. 1993. Catchment sensitivity to land use controls. Pp. 229–259 in *Landscape sensitivity,* edited by R. J. Allison and D. S. G. Thomas. West Sussex, UK: John Wiley & Sons.

Castelli, R. M., J. C. Chambers, and R. J. Tausch. 2000. Soil-plant relations along a soil-water gradient in Great Basin riparian meadows. *Wetlands* 20:251–266.

Costa, J. E. 1987. Hydraulics and basin morphometry of the largest flash floods in the conterminous United States. *Journal of Hydrology* 93:313–338.

Derbyshire, E., T. A. Dijkstra, A. Billard, T. Muxart, I. J. Smalley, and Y. -J. Li. 1993. Thresholds in a sensitive landscape: The Loess region of central China. Pp. 97--127 in *Landscape sensitivity,* edited by R. J. Allison and D. S. G. Thomas. West Sussex, UK: John Wiley & Sons.

Dohrenwend, J. C., A. Schell, and B. C. Moring. 1992. *Reconnaissance photogeologic map of young faults in the Millett 1 × 2 quadrangle, Nevada.* U.S. Geological Survey, Map MF-2176.

Downs, P. W., and K. J. Gregory. 1993. The sensitivity of river channels in the landscape system. Pp. 15–30 in *Landscape sensitivity,* edited by R. J. Allison and D. S. G. Thomas. West Sussex, UK: John Wiley & Sons.

Evans, R. 1993. Sensitivity of the British landscape to erosion. Pp. 189–210 in *Landscape sensitivity,* edited by R. J. Allison and D. S. G. Thomas. West Sussex, UK: John Wiley & Sons.

Gardiner, V. 1990. Drainage basin morphometry. Pp. 71–81 in *Geomorphological techniques,* edited by A. Goudie. London: Unwin Hyman.

Germanoski, D., and J. R. Miller. 1995. Geomorphic response to wildfire in an arid watershed, Crow Canyon, Nevada. *Physical Geography* 16:243–256.

Germanoski, D., C. Ryder, and J. R. Miller. 2001. Spatial variation of incision and deposition within a rapidly incising upland watershed, central Nevada. Pp. xi, 41–48 in *Sediment: Monitoring, modeling, and managing.* Proceedings of the 7th Interagency Sedimentation Conference. Vol. 2. Sponsored by the Subcommittee on Sedimentation. Mar. 25–29, 201. Reno, Nev.

Gerrard, J. 1993. Soil geomorphology, present dilemmas and future challenges. *Geomorphology* 7:61–84.

Hack, J. T. 1957. *Studies of longitudinal stream profiles in Virginia and Maryland.* U.S. Geological Survey Professional Paper 294-B.

Kleinhampl, F. J., and J. I. Ziony. 1985. *Geology of northern Nye County, Nevada.* Bulletin 99A. Nevada Bureau of Mines and Geology. University of Nevada, Reno.

Linsley, R. K., Jr., M. A. Kohler, and J. L. H. Paulhus. 1975. *Hydrology for engineers.* New York: McGraw-Hill.

McKee, E. H. 1976. *Geologic map of Austin Quadrangle, Lander County, Nevada.* Nevada Bureau of Mines and Geology, Map GQ-1307, 1:62,500 scale.

Miller, J. R., D. Germanoski, K. Waltman, J. R. Tausch, and J. C. Chambers. 2001. Influence of late Holocene hillslope processes and landforms on modern channel dynamics in upland watersheds of central Nevada. *Geomorphology* 38:373–391.

Patton, P. C. 1988. Drainage basin morphometry and floods. Pp. 51–64 in *Flood geomorphology*, edited by V. Baker, R. Kochel, and P. Patton. New York: John Wiley & Sons.

Patton, P. C., and V. R. Baker. 1976. Morphometry and floods in small drainage basins subject to diverse hydrogeomorphic controls. *Water Resources Research* 12:941–952.

Patton, P. C., and S. A. Schumm. 1975. Gully erosion, northern Colorado: A threshold phenomenon. *Geology* 3:88–89.

Perkins, J. S., and D. S. G. Thomas. 1993. Environmental responses and sensitivity to permanent cattle ranching, semi-arid western central Botswana. Pp. 273–286 in *Landscape sensitivity*, edited by R. J. Allison and D. S. G. Thomas. West Sussex, UK: John Wiley & Sons.

Roberts, N., and P. Barker. 1993. Landscape stability and biogeomorphic response to past and future climatic shifts in intertropical Africa. Pp. 65–82 in *Landscape sensitivity*, edited by R. J. Allison and D. S. G. Thomas. West Sussex, UK: John Wiley & Sons.

Schumm, S. A. 1977. *The fluvial system.* New York: John Wiley & Sons.

———. 1985. Explanation and extrapolation in geomorphology: Seven reasons for geologic uncertainty. *Japanese Geomorphological Union, Transactions* 6:1–18.

———. 1991. *To interpret the earth: Ten ways to be wrong.* Cambridge, UK: Cambridge University Press.

Shaw, P. A., and D. S. G. Thomas. 1993. Geomorphological processes, environmental change and landscape sensitivity in the Kalahari region of southern Africa. Pp. 83–96 in *Landscape sensitivity*, edited by R. J. Allison and D. S. G. Thomas. West Sussex, UK: John Wiley & Sons.

Stewart, J. H., and J. E. Carlson. 1976. *Geologic map of north-central Nevada.* Nevada Bureau of Mines and Geology, Map 50, 1:250,000 scale.

———. 1978. *Geology map of Nevada*, U.S. Geological Survey in cooperation with Nevada Bureau of Mines and Geology, 1:500,000.

Stewart, J. H., and E. H. McKee. 1977. *Geology and mineral deposits of Lander County, Nevada.* Bulletin 88. Nevada Bureau of Mines and Geology. University of Nevada, Reno.

Strahler, A. N. 1957. Quantitative analysis of watershed geomorphology. *American Geophysical Union Transactions* 38:913–920.

Chapter 5

Geomorphic and Hydrologic Controls on Surface and Subsurface Flow Regimes in Riparian Meadow Ecosystems

DAVID G. JEWETT, MARK L. LORD, JERRY R. MILLER, AND JEANNE C. CHAMBERS

Riparian ecosystems generally account for a limited amount (1–3 percent) of the total land area in arid regions (Naiman and Decamps 1997) such as the Great Basin of central Nevada. Nonetheless, they contain a disproportionally large percentage of the region's biodiversity (Saab and Groves 1992; Hubbard 1977). Water, from both surface and subsurface sources, is essential for the occurrence and persistence of Great Basin riparian areas and other similar riparian ecosystems in mountainous regions of the western United States. Changes in the abundance or the quality of water supporting meadow and other riparian ecosystems, whether the cause is natural or anthropogenic, have dramatic effects on ecosystem survival (National Research Council 2002). Therefore, the hydrologic regime is an important consideration for maintaining and restoring ecosystem pattern and process.

Riparian ecosystems in arid and semiarid regions have been greatly altered by both natural and anthropogenic disturbances. Changes in climate have had dramatic effects on the morphology of stream channels and adjacent riparian areas. Changes in vegetation patterns and geomorphic processes in central Nevada over the past 10,500 years have been used to track climate alterations (Chambers et al. 1998; chapters 2 and 3). A decrease in plant species numbers, significant hillslope erosion, formation of side-valley alluvial fans, and aggradation of valley floors occurred between 2500 and 1900 YBP during a dryer, cooler climate. During the past four hundred years, a warmer and wetter climate has resulted in increased plant species numbers, a decrease in hillslope erosion, and an increase in stream incision (Chambers et al. 1998; chapter 3).

Human activities such as flow regulation, surface and groundwater withdrawals, agricultural practices (grazing in particular), and recreational activities also affect the sustainability of riparian ecosystems (Patten 1998). In arid and semiarid regions there is a constant competition for water resources for drinking, livestock, and irrigation and for flows to sustain riparian ecosystems and aquatic species. Because water resources are limited, groundwater depletion and stream dewatering threaten many stream and riparian ecosystems (Ohmart et al. 1988; Cooper 1994; Stromberg et al. 1996). Arid region riparian corridors are a source of food, water, and shelter for domestic livestock. Overgrazing can deplete vegetation, destabilize stream banks, and increase sediment and nutrient loading (Kaufman and Krueger 1984; Feller 1998). Riparian corridors also are a focal point for recreational activities. The overuse of riparian recreation sites can degrade riparian ecosystems (Johnson and Carothers 1982), producing effects similar to overgrazing. In some cases, stream incision resulting from climatic shifts has been accelerated by these anthropogenic disturbances (Chambers et al. 1998; chapter 3).

Riparian vegetation patterns are directly controlled by geomorphic and hydrologic factors (Bendix and Hupp 2000; chapter 7). In riparian ecosystems, the groundwater regime is a primary control on plant species distributions (Allen-Diaz 1991; Stromberg et al. 1996; Castelli et al. 2000; Law et al. 2000; Martin and Chambers 2001). Stream incision on the order of less than 1 to 5 or more meters, brought on by climatic shifts and likely exacerbated by anthropogenic impacts, alters species composition and vegetation patterns. Stream entrenchment lowers the elevation of the stream surface that serves as local base level to the groundwater system. The local water table adjusts with a decrease in elevation that propagates through the shallow aquifer until a new equilibrium water level is established. The resulting decline in local groundwater tables can alter plant physiological and population processes (Martin and Chambers 2001, 2002) and community dynamics (Castelli et al. 2000). In the central Great Basin, ongoing stream incision results in a progressive decrease in the aerial extent of the riparian corridor and a loss of meadow complexes.

Successful maintenance and restoration of riparian ecosystems requires a good understanding of the surface and groundwater flow regimes and their interactions (Winter et al. 1998; Winter 1999; Woessner 2000; Jones and Mullholland 2000). An important component of the Great Basin Ecosystem Management research project focuses on increasing our knowledge of the hydrology of riparian meadow ecosystems in upland watersheds of central Nevada. Research conducted to date has (1) characterized surface and groundwater flow systems and the interactions between these hydrologic systems with respect to geomorphic-geologic setting, (2) documented the response of riparian hydrologic systems to geomorphic incision and evolution, and (3) evaluated relationships between surface water–groundwater systems and riparian vegetation patterns. Results from this research are being used to develop a framework for implementing scientifically sound riparian zone management and restoration strategies.

This chapter examines the geomorphic and hydrogeologic controls on riparian ecosystem hydrology of upland watersheds in the Great Basin of central Nevada. Two case studies are used to illustrate representative riparian hydrologic systems and the interactions between surface and subsurface waters. The relationships among vegetation and hydrology are examined and the impact of stream incision on hydrology and vegetation patterns is evaluated. A discussion of a framework for implementing scientifically sound management and restoration strategies from a hydrologic perspective concludes the chapter.

The Great Basin

The Great Basin consists of more than one hundred major hydrographic basins covering an area of approximately 500,000 square kilometers (Heath 1988; Mifflin 1988). The hydrology of the Great Basin is complex due to a combination of climate, basin and range structure, and internal drainage (Mifflin 1988). In central Nevada, the physiography of the Great Basin is dominated by southwest-northeast–trending fault-blocked ranges that are separated by intermontane basins. Regional groundwater flow systems in Nevada are common in large drainage basins and are characterized by regional boundary controls, discharge zones with minimal fluctuations in flow, and interbasin water transfers. Local flow regimes are characterized by (1) small drainage areas with relatively short flow paths; (2) boundary controls consisting of local lithologic, tectonic, and topographic features; (3) surface and groundwater discharges that vary greatly both seasonally and with storm events; and (4) little or no interbasin transfers of water (Maxey 1968).

Average annual precipitation in the Great Basin ranges from 100 to 500 millimeters and varies greatly throughout the region. Approximately 0–16 percent of the annual precipitation recharges local and regional groundwater systems (Maxey and Eakin 1949; Berger 2000; Nichols 2000). Winter snowpacks in mountainous areas are the primary source of groundwater recharge in higher elevation watersheds. Melting of these snowpacks provides for increased stream flows during the springtime. As spring melt-

lecrease, groundwater baseflow becomes the primary source of
staining stream flows. Perennial streams carry water from the
levations toward the large intermontane basins. Stream discharges
 throughout the summer and fall, except for occasional high flows
t on by large storm events. Surface water and groundwater ex-
occur along the entire length of the stream channel. Whether the
 a gaining, losing, or flow-through system (Winter et al. 1998;
'998) varies among valley segments and stream reaches and deal hydrogeologic and geomorphic controls. In the Great Basin,
ventually become losing systems as they cross and infiltrate
entering the internally drained intermontane basins,
is and groundwaters ultimately are evaporated or
in 1988).

nd Watersheds of Central Nevada

Whereas the physiographic setting of the Great Basin of central Nevada is
dominated by fault-block ranges separated by intermontane basins, the geographic setting of watersheds in upland areas is quite different. Upland watersheds typically are smaller in area, ranging from a few square kilometers
to more than 100 square kilometers, and generally occur at elevations ranging from 1,750 to 3,200 meters. Annual precipitation also is quite variable
and is orographically controlled. Higher elevations can receive precipitation in excess of 500 milliliters per year. Precipitation near the mouths of
upland basins is about 200 milliliters per year and approximately 60 percent of the precipitation occurs as snowfall during the winter months.

Stream systems in upland watersheds of central Nevada consist of axial
channels that are characterized by high gradients and gravel dominated
beds. Peak runoff is in response to snowmelt in late May or early June, but
localized convective storms during the summer can result in transient high
streamflow events. Average stream discharges range from 0.015 to 2.00
cubic meters per second (Hess and Bohman 1996). In general, stream flow
increases downvalley due to groundwater discharge to the channel. However, groundwater and surface water interactions are variable along the entire length of the axial stream system. At the mouths of the upland basins,
stream flow dissipates through evapotranspiration, recharges the groundwater system of the intermontane basin, or is diverted for irrigation purposes.

Vegetation at low to mid-elevations in upland watersheds consists of
Wyoming big sagebrush (*Artemisia tridentata wyomingensis*) communities
interspersed with Utah juniper (*Juniperus osteosperma*) and single leaf

pinyon (*Pinus monophylla*) woodlands. At higher elevations, mountain brush vegetation, including mountain big sagebrush (*Artemisia tridentata* spp. *vaseyana*) and limber pine (*P. flexilis*), dominate. Riparian vegetation generally consists of stringers of quaking aspen (*Populus tremuloides*), narrow leaf cottonwood (*P. angustifolia*), river birch (*Betula occidentalis*), and willows (*Salix*) (Weixelman et al. 1996). Vegetation patterns in riparian ecosystems are controlled by both geomorphic and hydrologic factors, in particular, the depth to groundwater is important (Allen-Diaz 1991; Stromberg et al. 1996; Castelli et al. 2000; Martin and Chambers 2001). Meadow vegetation types occur along distinct hydrologic gradients and include, from shallow to deep groundwater depths, Nebraska sedge (*Carex nebrascensis*) or wet meadow, mesic meadow, dry meadow, and basin big sagebrush (*Artemisia tridentata tridentata*) meadow types (Weixelman et al. 1996; Castelli et al. 2000).

Case Studies

Two case studies are used to highlight the importance of understanding the hydrogeologic and geomorphic setting in terms of maintaining and restoring meadow and other riparian ecosystems in the Great Basin of central Nevada. Both field sites are located in the Humboldt-Toiyabe National Forest and are believed to be representative of riparian corridors within upland watersheds.

Study Areas

The first site is located in the Big Creek watershed (39°20′30″N, 117°07′42″W) in the Toiyabe Range at an elevation of 2,300 meters, approximately 21 kilometers south of Austin, Nevada (see fig. 1.1, chapter 1). The Big Creek basin covers an area of approximately 23 square kilometers. More than 75 percent of the watershed is underlain by siliceous siltstones, quartzites, cherts, and, to a lesser degree, limestones of the Ordovician Valmey formation. At the higher elevations, the bedrock geology consists of Cambrian and Ordovician limestones and phyllites, as well as limited quantities (less than 1 percent) of weakly consolidated mudstones, siltstones, sandstones, and welded, ash-flow tuffs (McKee 1976; Stewart and Carlson 1976). Valley fill sediments and alluvial fan deposits overlay bedrock within the axial valley of Big Creek. The Big Creek site consists of a meadow complex located upstream of a valley constriction created by a side-valley alluvial fan that nearly extends across the entire width of the valley floor. A perennial channel, incised approximately 1 meter below the valley floor, traverses the meadow system and links upstream and downstream reaches of the basin. Several small, spring fed gullies also flow across the wet meadow to the incised axial channel.

The second site is located in Indian Creek valley (38°49'03"N, 117°30'02"W) in the headwaters of the Reese River basin; it is located about 60 kilometers south-southwest of the Big Creek site. The Indian Creek watershed, 86 square kilometers in area, is bound by the Shoshone and Toiyabe ranges. The elevation at the study site is about 2,225 meters. Volcanic bedrock, mostly rhyolitic tuff, underlies the basin, and Quaternary alluvium and colluvium cover the axial valley and gentle slopes (Kleinhampl and Ziony 1985). The Indian Creek valley is generally broad and supports about a half dozen meadow complexes that exhibit varying degrees of channel incision. Upstream, channels are absent or discontinuous with wet and mesic meadows that cover most of the valley floor. In contrast, downstream, wet and mesic meadows are confined to the floor of incised trenches where the groundwater table has dropped. The progressive stages of entrenchment of the meadow complexes have afforded an important opportunity to study the interdependent responses of hydrologic, geomorphic, and vegetational components of the valley system.

Methods

Field reconnaissance and mapping was conducted to describe the geomorphic and stratigraphic settings of the Big Creek and Indian Creek field sites. Topographic and geomorphic features, such as stream channels, terraces, valley floors, alluvial fans, and hillslopes were surveyed at both sites. Stratigraphic sections, where available, also were used to interpret the geomorphic and hydrogeologic settings. In addition to site specific data, field reconnaissance data from other riparian meadow complexes in the Toiyabe, Toquima, and Monitor Ranges of central Nevada, with similar valley floor morphologies, also have been used to clarify hydrogeomorphic relations.

Shallow groundwater flow at the Big Creek and Indian Creek field sites was characterized based on hydraulic head (water level) data collected from a large number of piezometers and observation wells. Piezometers are small-diameter wells with a short screened section used to measure the total energy of water at a point in an aquifer, also known as hydraulic head (Fetter 2001). Observation wells generally have a longer screened section and are installed so that the water table lies within the screened section.

Groundwater flows from higher to lower hydraulic head values. Thus, hydraulic head data from the piezometers and wells were used to construct water table maps and profiles.

Heavy equipment could not be used in the sensitive meadow environments, therefore all wells were installed by hand. In general, the piezometers were constructed of 1.9-centimeter (0.75 inch) thin-walled, electrical conduit. The lower 0.5 meter of the conduit was slotted with a power hacksaw or drilled with small holes to provide a "screened" section. The lower end of the conduit was then plugged with a machined drive point or carriage bolt and the piezometer was driven into the ground using a fencepost hammer. Once the desired depth was obtained, between 0.5 and 6 meters, the drive point or bolt was driven out of the end of the conduit to facilitate water flow into the pipe.

At several locations at both study sites, nested piezometers were constructed. Nested piezometers are closely spaced piezometers open at different depths in the subsurface that provide data on the vertical hydraulic gradients in the area of the nests (Fetter 2001). The nested piezometers in this study consisted of two piezometers located in close proximity, one installed to a shallow depth (0.6–2.6 meters) and the other to a greater depth (2.3–5.3 meters) depending on the depth to groundwater. The conduit used for nested piezometers was not slotted, only open-ended.

Locations and elevations of wells and topographic and geomorphic features were mapped using a Leica Total Station. All elevations were determined relative to a site benchmark. Reliability of survey data, particularly groundwater well elevations and locations, was assessed by repeating the survey from separate locations (where trees and other vegetation allowed a view of the sites). In general, the validation measurements were within 1 centimeter of the original survey data. All maps are based on these survey data. Map accuracy was determined by visual inspection of cartographic outputs in the field.

The Big Creek Site

GEOMORPHOLOGY

The Big Creek watershed is characterized by an integrated stream system that is incised into a low-relief, narrow (less than 150 meters wide) valley floor. The Big Creek study area is bounded to the north (downstream) by two coalescing side-valley alluvial fans and to the south by another side-valley fan deposit (fig. 5.1). Steep bedrock hillslopes, covered with a thin ve-

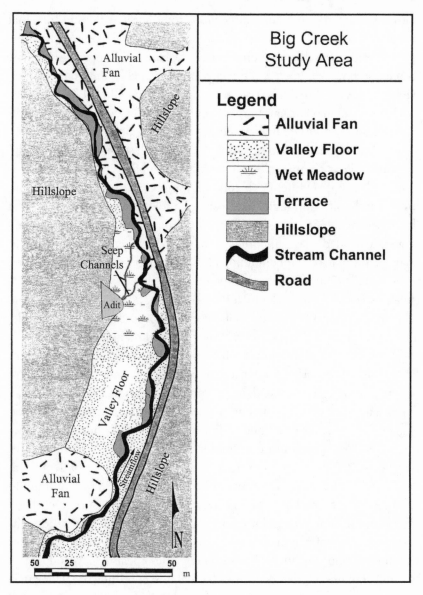

FIGURE 5.1. Landform map of the Big Creek study area. The downstream direction is to the north.

neer of colluvium, rise to the east and west. The meadow complex at the Big Creek site is located immediately upvalley of a large alluvial fan that has built away from a side-valley tributary, a feature common to many of the meadow ecosystems in central Nevada. These side-valley alluvial fans, in most cases, are composed of a single, geomorphic surface (Miller et al.

2001). Radial fan profiles measured along these surfaces demonstrate that during aggradational events, many of these fans extended completely across the valley floor with deposits impinging on the base of the opposing hillslope. At present, however, the toe of the fans are truncated by the axial channel, which is now confined between hillslope and distal alluvial fan deposits. Where truncation of the fans has not occurred, the surface of the fan deposits grade smoothly onto the highest surface preserved along the valley floor.

Field reconnaissance suggests that the valley fill stratigraphy within the Big Creek site is comparable to that observed in other upland watersheds of central Nevada (Miller et al. 2001). Exposures of the fan deposits were limited at the site, although an exposure 2.5 meters high was examined at the downstream end of the meadow. The exposed fan materials consisted of weakly bedded, clast supported gravels containing a loamy, medium to coarse sand matrix. Failed attempts to install hand-augured monitoring wells, and the difficulties of installing the piezometers, suggest that the up-stream fan consists of similar materials. Based on the regional stratigraphic relations (chapter 3), it is likely that these fan deposits interfinger with fine- and coarse-grained alluvial facies along the distal margins of the fans.

Within the meadow complex, the upper 0 to 5 centimeters of the sur-ficial materials consists of a dense mat of roots and undecomposed plant materials forming a peat-like deposit. Immediately underlying the peat-like deposits are dark-colored, organic-rich sediments that contain higher per-centages of mineral matter, and which are dominated by fine-grained sed-iments. Auger cuttings from exploratory boreholes suggest that this layer may exceed 0.5 to 1 meter in thickness. The sedimentology of deposits at depths below the materials observed within the cores is more difficult to ascertain. However, at most other sites where similar exposures have been observed, the deposits are dominated by horizontally bedded sands, silts and clays, some of which contain abundant plant macrofossils and roots which form thin (less than 10 centimeters thick) peatlike layers. Locally, this unit contains interbedded paleochannel fills, consisting of a wide range of sediment sizes, that are inset into (and abut) the finer-grained valley fill. It is not uncommon for these paleochannel fills to contain lenses of coarse-grained, clast-supported gravel that completely fill the channel and spill onto the valley floor.

HYDROGEOLOGY

A total of fifty-four piezometers were installed at the Big Creek site, in-cluding three sets of nested piezometers. The piezometer locations were distributed across the site in order to provide adequate spatial resolution of the water table surface (fig. 5.2).

Piezometer data provide a glimpse of the hydroperiod, that is the frequency and amplitude of water-level fluctuations (Winter et al. 1998), for different areas of the Big Creek site. Figure 5.3 illustrates the depth to the water table (average depth ± 1 SD) during the 1997–1999 field seasons for piezometers located in the upper (southern), mid, and lower (northern) regions of the site. Depth to water data were not collected throughout the entire year due to difficulties in accessing the site during the winter season. Monthly precipitation data and stream stage data also are provided in figure 5.3.

The depth to water is greatest in the upper (southern) end of the Big Creek site and water depths decrease toward the lower (northern) end of the site. Surface water and groundwater levels tend to be at their highest following the annual snowmelt and spring rains and the depths to water across the site decrease throughout the remainder of the year. The annual change in water table depths is greatest in the upper region and least in the lower region. Variations in water table depths also are correlated with longitudinal position (as indicated by standard deviation results, fig. 5.3). Surface water and groundwater levels drop throughout the rest of the year except during occasional storm events. For example, during two storm events in early September 1998 when approximately 15 and 31 millimeters of rain, respectively, fell over four days, water levels in the stream and in the piezometers located in the upper region of the site increased while the water levels in the other piezometers continued to decrease. This suggests that groundwater in the upper region of the Big Creek site is influenced by short-term precipitation events that also increase stream stage or increase groundwater input originating from the side-valley alluvial fan deposits.

Hydraulic head data from the shallow piezometers have been contoured (SURFER, Golden Software 1995) to provide maps of the water table surface for the Big Creek site during the wet and dry seasons (figs. 5.4A and B). The contour lines represent lines of equal hydraulic head or potential energy (also called *equipotentials*) relative to an arbitrary datum. Groundwater flows from higher to lower equipotentials (arrows on figs. 5.4A and B).

Figure 5.4A presents the water table surface at the Big Creek site during a wet season monitoring event (data were collected Jun. 26, 1998). Groundwater enters the study area through subsurface flow in the valley fill. Water supplied from Big Creek and the side-valley alluvial fan also influence groundwater flow in the upper (southernmost) portion of the site. The water table surface indicates that the upper reach of Big Creek is a *los-*

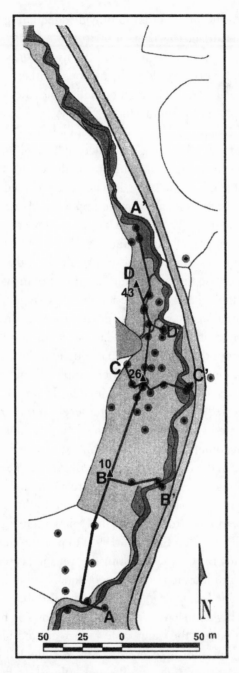

FIGURE 5.2. Piezometer locations and orientation of cross sections for water table profiles at the Big Creek site (● denotes piezometer location; ▲ denotes nested piezometer location).

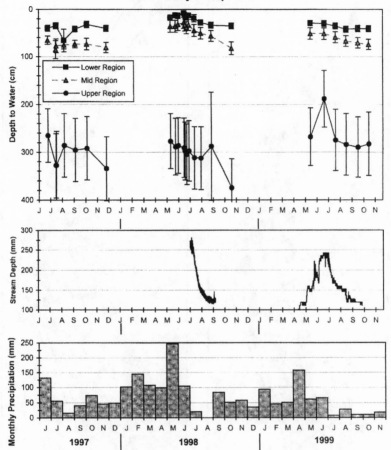

FIGURE 5.3. Mean depth to the water table (±1 SD) in piezometers located in the upper (southern), middle, and lower (northern) areas of the Big Creek site, stream depth (data courtesy of Michael Amacher, USDA Forest Service), and monthly precipitation from the Big Creek summit weather station (data courtesy of USDA Natural Resources Conservation Service).

ing stream (diverging groundwater flow lines) providing water to the groundwater system. The alluvial fan deposits in this area also provide a subsurface source of water input to the site. Groundwater moves down-gradient, to the north, through the midsection of the site. Big Creek then becomes a gaining stream, receiving baseflow from the subsurface, as it traverses the mid to lower reaches of the site as indicated by the concave pattern of equipotential lines (converging groundwater flow). The wet meadow occurs where the water table surface intersects the surface to-

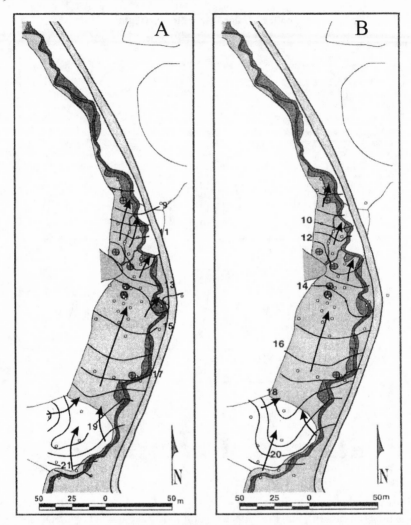

FIGURE 5.4. Big Creek water table surface: (A) wet season monitoring event (26 June 1998), and (B) dry season monitoring event (18 August 1997). Groundwater flow from higher to lower equipotentials (contours) is indicated by arrows.

pography. Soils are saturated in this area of the site during the wet season and standing water and surface runoff are readily apparent.

Similar groundwater flow patterns are evident at the Big Creek site during the dry season as illustrated by the water table map for a dry season monitoring event (fig. 5.4B, data collected Aug. 18, 1997). Inputs from the stream and the alluvial fan deposits still occur, but they are not as pronounced as evidenced by the increasing depth to water and the decreasing hydraulic gradient. Groundwater flows down gradient where the depth to water is slightly deeper through the midsection of the site. Groundwater discharge to Big Creek still occurs along the lower reaches of the site. Field observations indicate that the area of saturated soils has decreased slightly. Surface runoff still occurs, but discharge has decreased compared to wet season observations.

To better illustrate hydraulic gradients and differences in groundwater levels between wet and dry seasons, water table profiles have been produced for four cross sections through the site (fig. 5.5A–D). One cross section is oriented parallel to the axial valley (A-A', fig. 5.5A), while the remaining profiles are oriented perpendicular to the axial valley. The locations of the cross sections are indicated on figure 5.2. The data used to create these water table profiles are the same used to generate figure 5.4A and B.

Cross section A-A' (fig. 5.5A) is a south-to-north, longitudinal water table profile along the axial valley. This cross section is oriented parallel to the principal direction of groundwater flow through the site. The depth to water is greatest under the alluvial fan in the upstream area of the site. The water table approaches the ground surface farther downvalley and is at or near the ground surface where groundwater discharges to the meadow. The hydraulic gradient is steepest in the southernmost portion of the site where the downvalley subsurface flow in the valley fill deposits is augmented by inputs from the stream and the alluvial fan. The hydraulic gradient decreases across the lower portion of the fan and dry meadow but increases again where it intersects the ground surface. The wet season water table occurs at a shallower depth throughout the site compared to the dry season water table. The difference in depth between these two monitoring events is greatest in the upper (southern) region and least in the lower region near groundwater discharge zone.

Cross section B-B' (fig. 5.5B), a west-to-east cross section across the upper region of the site, illustrates marked differences between wet and dry season groundwater flow patterns. During the wet season, Big Creek is a losing stream, recharging the subsurface system. Even though the stage of Big Creek has declined from the spring 1998 high-flow events, bank storage in the vicinity of the stream remains high. With time, the water table elevation in the vicinity of Big Creek declines, approaching equilibrium with the water level in the stream. During the dry season, the water table flattens out along this profile indicating that lateral flow from the stream is minimal and the majority of groundwater flow is in a downvalley direction (fig. 5.5A). Cross section C-C' (fig. 5.5C) is a west-to-east profile through the midsection of the site. During the wet season, the water table intersects the

A

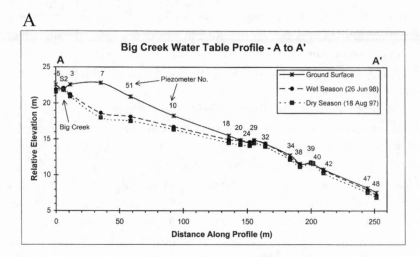

B

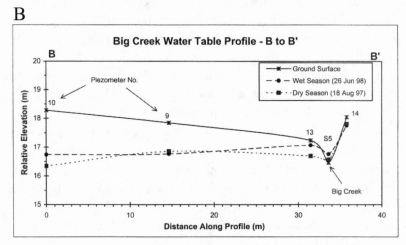

FIGURE 5.5. Water table profiles for the wet and dry season monitoring events along four cross sections at the Big Creek site: (A) longitudinal profile along axial valley, cross section A–A′, and three transverse profiles; (B) cross section B–B′; (C) cross section C–C′; and (D) cross section D–D′. See figure 5.2 for location and orientation of cross sections.

ground surface in the vicinity of piezometer no. 24, producing a seasonal discharge of groundwater. During the dry season, groundwater discharge to the surface is absent and the water table resides 30–40 centimeters below the ground surface at this location. Field reconnaissance during the dry season also reveals the decrease in the areal extent of saturated soils. Along the lower reaches, Big Creek is a gaining stream, receiving subsurface flow from the groundwater system during both the wet and dry seasons. The dis-

C

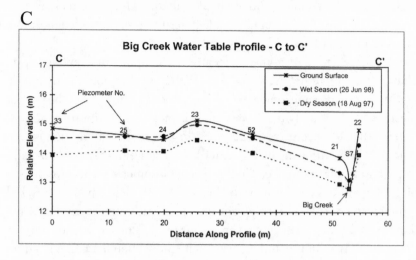

D

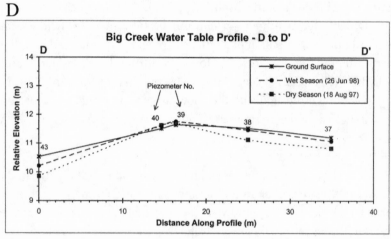

charge to Big Creek is slightly greater during the wet season as indicated by the slightly greater hydraulic gradient. Cross section D-D' (fig. 5.5D) crosses the lower region of the site along a southeast-to-northwest trend. The water table profiles illustrated in this cross section indicate that groundwater discharges to the surface in the center of the lower region (in and around piezometer nos. 39 and 40) during both the wet and dry seasons. This observation is corroborated by the presence of active seeps in this area of the site. Groundwater flow is from this area of discharge toward Big Creek to the east and to the north (figs. 5.2 and 5.5D), with the downvalley hydraulic gradient greater than the lateral gradient toward the stream.

Groundwater recharge and discharge patterns at the Big Creek site also are observed in vertical hydraulic gradient results from the nested piezometers. Piezometers 10s and 10d, 26s and 26d, and 43s and 43d (s = shallow; d = deep) are nested piezometers located in the upper, mid, and lower regions of the site, respectively (fig. 5.2). In the upper region, the hydraulic head in the shallow piezometer (10s) is greater than the head in the deeper piezometer (10d) and groundwater flow is therefore downward. Groundwater recharge is occurring in the upper region of the site throughout the year (or at least during the field season when measurements were collected). In contrast, the hydraulic head in the deeper piezometer in the lower region of the site (43d) is greater than the head in the shallower piezometer and groundwater flow is in an upward direction. Although this vertical hydraulic gradient occasionally reverses (downward flow), data indicate that the gradient is predominantly upward. The midsection of the site is a transition zone with downward hydraulic gradients during some years and upward gradients during other years. During the 1997 and 1999 field seasons, water level measurements at piezometers 26s and 26d indicate that a slight downward gradient was present. However, during the 1998 field season, the hydraulic gradient at piezometers 26s and 26d was upward. This upward gradient is associated with the above-average precipitation received in the area (fig. 5.3) and corresponding higher water table elevations (fig. 5.4A).

The Indian Creek Site

Indian Creek valley contains several meadow complexes that have undergone various degrees of channel incision. The entrenchment of the valley floor is largely driven by groundwater sapping, that is, bank failure triggered by undercutting caused by groundwater seeps. Incision has caused a lowering of the groundwater table and a vertical displacement of wet meadows from the valley floor down to the trench floor. A study area was chosen in Indian Creek valley to include different stages of incision based upon the geomorphic, hydrologic, and vegetational characteristics. A substitution of space (i.e., reach position) for time is used to predict the consequences of future incision in Indian Creek valley; this is a commonly used approach in geomorphology and has also been used to predict the vegetational consequences of groundwater decline (e.g., Stromberg et al. 1996).

GEOMORPHOLOGY

The Indian Creek study site is centrally located within its watershed and is coincident in location to where the basin width decreases from about 10 kilometers to 4 kilometers in the downstream direction (fig. 5.6). Upstream from this area, broad, gently sloping alluvial fan surfaces flank the axial valley and outcrops of bedrock are limited to high elevations along the drainage divides. The large alluvial fan surfaces are generally bouldery at the surface and are no longer active. Smaller and younger alluvial fans are common adjacent to the valley floor (fig. 5.6). At the downstream end of the study site, basalt associated with Black Mountain is exposed or covered with a thin veneer of colluvium.

The valley floor reaches widths of hundreds of meters at the upstream end of the basin, but narrows to tens of meters downstream. Upstream, about a half-dozen broad meadow complexes cover the valley floor, which are only dissected by shallow, discontinuous channels. These trench-shaped channels are generally low relief (less than 50 centimeters), have steep-walled headcuts, and, in some places, terminate in shallow fanlike features composed of sediment deposited on the valley floor. As the valley narrows downstream, in the study site a continuous trench, about 2.5 kilometers in length, has been incised into the valley floor (fig. 5.6). Several subparallel trenches are tributary to the main trench in the upstream portion of the study area. The main trench is rectangular in shape, up to 4 meters deep, and 30 meters wide; the trench size decreases downstream. Trench walls have steep sides and are blanketed with fresh slump blocks. The trench itself is not a stream channel. The trenches share some characteristics of arroyos, most notably the vertical walls formed by erosion into axial valley alluvium and the presence of a stream channel on the trench floor. However, arroyos are typically larger-scale and more continuous erosional features with the incision generally caused by ephemeral stream channels (Bull 1997; Elliott et al. 1999). Many of the trenches in Indian Creek valley are gullies in that they are discontinuous and formed where there were no prior channels (Schumm 1999).

A small stream channel, Indian Creek, is located on the trench floor and is 0.5 to 2 meters wide. Upstream within the study area, the stream bed is fine grained, and supports perennial flow. Downstream, flow is intermittent or ephemeral and the stream bed is cobbly. The stream channel form also changes in the downstream direction from well-defined, deep, wide pools connected by small runs to a wide, shallow channel. One or two low-relief (tens of centimeters) stream terraces are present along some reaches of the trench floor.

Preliminary stratigraphic analysis of exposed sections in Indian Creek valley indicate four distinct depositional units (Lord et al. in review). The

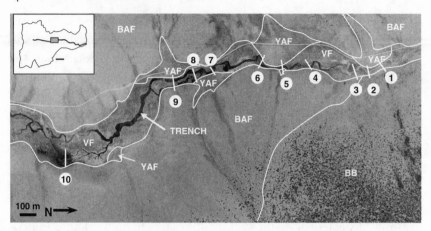

FIGURE 5.6. Vertical aerial photograph of Indian Creek valley showing the study area, geomorphic units, and locations of study transects (numbered lines). Geomorphic units are VF, valley floor; TRENCH, incised trench within valley floor; YAF, young alluvial fan surface; BAF, older, bouldery alluvial fan; and BB, basalt bedrock near surface associated with Black Mountain (northeast corner of photo). Groundwater measurements and vegetative surveys were made along transects. Index map shows watershed shape and study location within the basin.

basal unit, which is dated with radiocarbon to 2350 ± 70 YPB, is interpreted to correlate with a regional period of aggradation between 2580 and 1920 YBP (Miller et al. 2001; chapter 3). The upper stratigraphic units at Indian Creek include channel fill deposits and discontinuous, horizontally laminated sediment thought to represent channel cutting, channel filling, and fan deposition by unchannelized flow. A pervasive, organic-rich horizon caps the section throughout the upstream portion of the study site and is interpreted as a wet meadow paleosol (a buried soil horizon).

HYDROGEOLOGY

Ten transects were established in four reaches of Indian Creek valley with each reach representative of a different stage of valley floor incision (fig. 5.6). The emphasis of this discussion is on the three reaches that are in the incised section of Indian Creek valley with the depth of incision greatest upstream. The fourth reach (at transect 10) is in a broad, meadow complex just upstream of headcuts, which form the terminus of the trench system (fig. 5.6). Along each transect, groundwater wells were installed on different geomorphic surfaces (i.e., valley floor, trench floor, trench floor terrace) on both sides of the stream. Vegetational composition was determined at each well site. Four or five wells were installed along most transects. In each of the lower three reaches, one transect was instrumented in more detail (transects 3, 5, and 8). Along these transects, nested piezo-meters were installed on both sides of the trench floor and observation wells were installed nearby up- and downstream of the main transect line. This design permitted a local, three-dimensional view of groundwater flow patterns. A total of sixty groundwater wells were installed at the Indian Creek site. As with the Big Creek site, groundwater levels were measured at least once per month from May to November. Three years of ground-water level data have been collected at Indian Creek.

The depth to groundwater strongly varies with geomorphic position and reach (table 5.1). In the incised reaches of Indian Creek valley, the depth to the water table is greater from the valley floor, typically 2 to 4 meters, than the floor of the incised trench. In contrast, in the unincised meadow complex upstream (transect 10), the mean depth to water table is less than 1 meter on the upgradient side of a trench headcut (fig. 5.6). The ground-water table is 2 to 3 meters lower on the downgradient side of the trench. Along the floor of the incised trench, the depth to groundwater increases downstream, ranging from near or at the ground surface upstream to well over 1 meter deep in the area downstream.

The position of the groundwater table is significantly more spatially and temporally variable downstream than upstream. For example, over the three-year monitoring period (twenty water-level readings) the water table elevation downstream (transect 3) has ranged about 2.5 meters while up-stream (transect 8) the water table elevation has ranged less than 0.3 me-ters. On a given date, the water table elevation along a transect typically varies between 0.1 to 0.2 meter at transect 8 and about 0.4 meter at tran-sect 3. Seasonal variations in the groundwater table become increasingly pronounced with distance downstream from the trench headcuts (fig. 5.7). In the upstream reach, the level of the groundwater table elevation is very consistent whereas the downstream reach shows distinct seasonal patterns in the groundwater table with about 2 meters of drop over the growing season.

The interaction of groundwater and stream water can be evaluated by examination of water table maps and nested piezometer data. A water table map of the upstream incised study reach, covering transects 7 to 9, indi-cates a gaining stream (converging groundwater flow) at the upstream end of this reach and a losing stream at the downstream end (fig. 5.8). The set of nested piezometers on the west side of transect 8 corroborates this ob-servation by exhibiting an upward groundwater flow gradient (table 5.1). Elsewhere in the incised portion of the study area, Indian Creek is mostly losing, but the conditions vary with space and time. For example, although

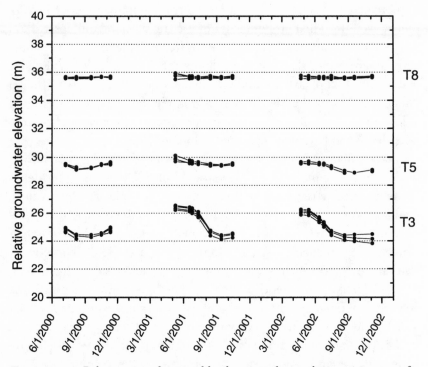

FIGURE 5.7. Relative groundwater table elevations during the growing season for a cross-sectional set of wells at upstream (T8), midstream (T5), and downstream (T3) sections. The groundwater table at the downstream section exhibits the greatest spatial and seasonal variability. Individual lines are temporal trends for an individual well.

the temporal average vertical hydraulic gradients at transect 5 (table 5.1) indicate upward flow, the gradients are very low and are negative (i.e., downward) as often as positive. The downstream incised reach (transects 1 to 3) is consistently losing with a relatively steep downward gradient.

Groundwater and stream interaction can be further assessed by analysis of the electrical conductivity of water samples. On three dates, groundwater wells and stream water were sampled for conductivity at transects 1 to 9. The conductivity of the groundwater at the upstream and midstream reaches (means ± SD: 515 ± 106 and 532 ± 129 microsiemens per centimeter, respectively) were significantly higher than the downstream reach and the stream water (means 409 ± 50 and 445 ± 2.8 microsiemens per centimeter, respectively). The very low variability in the stream water conductivity indicates little influence of groundwater on stream water chemistry over the study reach. The relatively low conductivity of the groundwater downstream is consistent with either the stream water being the

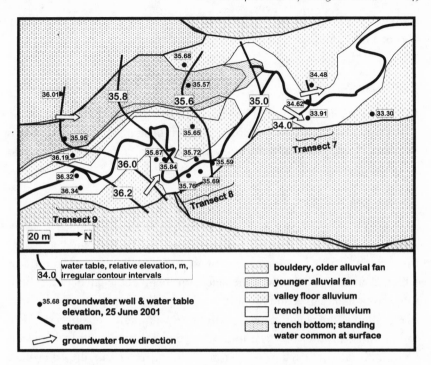

FIGURE 5.8. Groundwater table, well locations, and geomorphic units of the upstream section. The stream trends from gaining to losing in the downstream direction over the reach shown; consistent with lateral trends shown, nested piezometers in this reach most commonly exhibit an upward groundwater flow gradient.

TABLE 5.1.

Mean (±1 SD) depth to water table (meters) and the hydraulic gradients in the different reaches of Indian Creek valley

Locations	Depth to water table from valley floor	Depth to water table from trench floor	Water table gradient, down valley	Vertical hydraulic gradient, east side of stream[*]	Vertical hydraulic gradient, west side of stream
Downstream (T3)	3.0 ± 0.84	1.21 ± 0.83	0.0038	−0.081	no data
Midstream (T5)	2.9 ± 0.21	0.57 ± 0.19	0.010	0.0040	0.0099
Upstream (T8)	2.4 ± 0.13	0.15 ± 0.065	0.011	−0.074	0.013
Broad Meadow (T10)	0.96 ± 0.32	NA	no data	no data	no data

[*]A negative value indicates a downward groundwater flow component; a positive value indicates an upward flow component

source of the groundwater (i.e., a losing stream) or the groundwater being part of a different flow system than upstream reaches.

Synthesis and Review

Detailed longitudinal and cross-valley surveys conducted as part of the Great Basin Ecosystem Management Project illustrate that the alluvial fans can have a significant influence on channel morphology, hydrology, and meadow development (chapter 3).

Geomorphic Setting

At many sites, stream entrenchment tends to be most pronounced at or immediately upstream of the side-valley alluvial fans where knickpoints on the order of 0.5 to 2 meters in height are developed in fan sediments, or within axial valley alluvium immediately upstream of the fan deposits. Both the active channel bed and the surface of the valley floor dramatically drop in elevation as the channel traverses the side-valley fans, creating a stepped, longitudinal profile. For example, at several locations within Kingston Canyon, located across the divide from the Big Creek basin, the valley floor drops by more than 3 meters in elevation as the stream traverses the toe of a fan that constricts the valley width. These abrupt changes in valley floor elevation result in significant variations in channel gradients. Reaches upstream of the fans exhibit relatively low gradients. Slopes increase significantly as the channel traverses the fan deposits, and then systematically decrease farther downstream, forming a concave-up longitudinal profile between successive fan reaches. Thus, the depth of incision, and the gradients of the channel and valley floors, vary along the riparian corridor and are directly related to the geomorphic position of the reach relative to the location of the larger, side-valley fan complexes.

Changes in channel gradients, and a stepped, longitudinal profile, do not occur at all of the alluvial fans found within the upland watersheds. This is clearly illustrated at the Big Creek site where alluvial fans constrict the valley at the downstream end of the meadow complex, but significant changes in channel gradients do not occur. The occurrence of meadows upstream of side-valley alluvial fans, where stepped, longitudinal profiles do and do not occur, suggest a genetic linkage between the meadows and fan deposits.

The origin of the meadow complexes in the Indian Creek valley is not clearly related to the position of the side-valley alluvial fans. Lord et al. (in review) suggest that the large basin width, coarse-grained alluvial fans, and low-angle slopes in the upper portion of the basin provide ideal conditions for groundwater recharge (fig. 5.6). Downstream, the basin and valley width narrow, the bedrock is different, and the sediment cover is thin. These conditions likely cause groundwater flow to converge upgradient of the "constriction," affording a steady supply of groundwater. Although the Indian Creek valley meadow complexes are not related to side-valley alluvial fans, the Indian Creek basin geology and topography do serve a similar hydrogeologic function in that this setting causes a convergence and concentration of upgradient groundwater flow. Regardless of the origin of individual riparian meadows in the upland basins, it is evident that the sum of the geologic and topographic conditions associated with the setting must be conducive to supplying ample and steady groundwater—most likely as a discharge area of a large-scale groundwater flow system.

Hydrogeologic Setting

Physical hydrogeology findings for the Big Creek site indicate that the predominant component of subsurface flow is in the downvalley direction within the valley fill deposits. The upper (southern) region of the Big Creek site is a groundwater recharge zone as indicated by the vertically downward hydraulic gradients. Sources of water for this recharge zone include subsurface inputs from the upgradient valley fill sediments, subsurface inputs from side-valley alluvial fan deposits, and surface water inputs from Big Creek (figs. 5.4A and B). The influence of the side-valley alluvial fan is more pronounced during the wet season than during the dry season, which is likely due to increased flows associated with spring snowmelt and runoff. Along its upper reach, Big Creek is a losing stream during both the wet and dry seasons.

Farther downstream, Big Creek recharges the subsurface system during periods of higher stream flow. However, the hydraulic gradient between the streambed and the shallow aquifer, and thus the amount of recharge, decreases as stream stage declines over time. Groundwater flow is roughly parallel to the trend of the axial valley, through the midsection of the site. The lower region of the site, in the vicinity of the wet meadow area, is a groundwater discharge zone. The nested piezometers in this area indicate the presence of an upward vertical hydraulic gradient. Groundwater discharges in seeps throughout the wet meadow and flows along the ground surface to Big Creek. Groundwater also discharges directly to the stream as baseflow. Some groundwater likely continues moving downgradient in the valley fill sediments.

At the Indian Creek site, the characteristics of the groundwater system are best explained by a near-constant supply of groundwater at or above the upstream end of the study site. This is indicated by the steady groundwater table levels found in transects 7 to 9 (figs. 5.6 and 5.7). Furthermore, the absence of seasonal variation in the groundwater table indicates a discharge zone of deep groundwater flow system, which is not depleted by evapotranspiration (Mifflin 1988). These observations are similar to those of Hill (1990) in a study of groundwater discharge in an upland meadow in northwestern Nevada. Hill instrumented a headwaters meadow dissected by a gully with wells and piezometers. Based on physical and chemical hydrologic data, he interpreted the dominant groundwater source to be deeper discharge from fractured rhyolitic bedrock rather than from alluvium and colluvium.

Downstream in the Indian Creek site, the variability of the groundwater table increases and the stream becomes consistently losing. In the middle reach, transects 4 to 6, reversals in vertical groundwater flow directions are common through time. In this area, the lateral, downvalley, and vertical hydraulic gradients are low. Therefore, subtle shifts in the balance of lateral versus downvalley groundwater inputs cause changes in the interaction of groundwater and stream water. Temporal shifts in the location of gaining and losing reaches of streams is common in the Great Basin (Mifflin 1988). In the downstream reach (transects 1 to 3), the generally dry stream channel, variable groundwater table elevation, and low hydraulic gradient likely reflect an increase in the permeability of subsurface materials caused by the change in geology and the large distance from the upstream, steady groundwater source.

Groundwater discharge in the wet meadow areas is likely due to (1) a decrease in the hydraulic conductivity and connectivity of valley fill deposits at or immediately downstream of the meadows, (2) constrictions in the width and thickness of the valley fill aquifer materials, or (3) a combination of the two. Either of these changes would be accompanied by an increase in the hydraulic gradient near the wet meadow such as the one observed in the longitudinal water table profile (fig. 5.5A) of the Big Creek site. A dramatic decrease in the ability of the sediments to transmit water could result in groundwater discharge at the surface in order to maintain hydrologic continuity. Geomorphic and stratigraphic evidence indicate that the valley fill deposits are highly complex and spatially variable. This suggests that the hydraulic conductivity and lateral continuity of these deposits also are highly variable and may, in part, be responsible for the presence of wet meadows upstream of the alluvial fans.

A decrease in the cross-sectional area available for subsurface flow, due to a constriction in the width or thickness of the valley fill sediments, also could cause groundwater discharge. At the Big Creek site, the geometry of the valley fill aquifer changes near the downstream alluvial fans (fig. 5.1). The width of the valley narrows in the downstream direction, thereby reducing the cross-sectional area of the valley fill sediments. In order to maintain hydrologic continuity, groundwater discharges to the wet meadow. Groundwater discharge and wet meadow formation could also be due to a vertical constriction created by the thinning of valley fill sediments over a subsurface bedrock high.

Given that most meadows occur upstream of side-valley alluvial fans, and that groundwater discharge may be associated with a reduction in the cross-sectional area of the aquifer as a result of bedrock highs, there appears to be a genetic link between the side-valley fans and evolution of the bedrock topography underlying the valley floor. Miller et al. (2001) found that the side-valley fans are composed of sediments from numerous depositional sequences, the most recent occurring between approximately 2500 and 1900 YBP. Thus, fan deposition has been an episodic process throughout the Holocene. It is possible that prior to valley aggradation, scour of the bedrock in the vicinity of the side-valley fans was limited by the delivery and deposition of coarse-grained sediments from the fan basins. Put differently, the fan deposits protected the underlying bedrock from erosion, creating localized highs in the longitudinal configuration of the bedrock topography beneath the existing valley floor.

Smaller-scale groundwater and surface water interactions also likely occur within and in close proximity to the stream channels. Numerous researchers have reported on the importance of groundwater–surface water mixing in the hyporheic zone (Stanford and Ward 1988; Valett et al. 1990; Harvey and Bencala 1993; Winter et al. 1998; Wroblicky et al. 1998). These studies show that surface and subsurface exchanges of water occur within pool and riffle sequences, both vertically through the streambed and laterally through the stream banks. These smaller-scale interactions also are important to the development and the persistence of the riparian ecosystems.

Vegetation and Groundwater Relations

The primary controlling factor of plant species distribution in arid and semiarid riparian ecosystems is the availability of water from in-stream or groundwater sources (Allen-Diaz 1991; Stromberg et al. 1996; Castelli et al. 2000; Law et al. 2000; Martin and Chambers 2001). This relationship between vegetation and depth to groundwater is evident at both of the study areas. A brief review of this vegetation-groundwater level relationship is provided here. Chapter 7 presents a more indepth discussion of the relationships between geomorphology, hydrology, and vegetation patterns.

At the Big Creek site, a hydrologic gradient exists along the trend of the axial valley. The greatest groundwater depths occur beneath the upstream alluvial fan (figs. 5.3 and 5.5A). The depth to the water table gradually decreases in a downstream direction, with groundwater eventually discharging at the surface in the wet meadow area. Castelli et al. (2000) studied the spatial and temporal relations between vegetation, soil, and hydrologic gradient at two Great Basin riparian meadows, including the Big Creek site. Their analysis showed that study variables associated with plant species composition were strongly related to water table depth.

Vegetation at the Big Creek site was classified into four ecological site types, including wet meadow, mesic meadow, dry meadow, and basin big sagebrush–basin wildrye (Weixelman et al. 1996; Castelli et al. 2000). These four meadow types are closely related to the depth to the water table and the relative variation in the water level. The wet meadow exists where the water table is within 30 centimeters of the ground surface and seasonal variations in the water table elevation are minimal. Dominant plant species in the wet meadow are Nebraska sedge (*Carex nebrascensis*) and Baltic rush (*Juncus balticus*). The mesic meadow type is located where the water table ranges from 30 to 80 centimeters below the ground surface and consists of Kentucky bluegrass (*Poa pratensis*), Baltic rush (*J. balticus*), and creeping bent grass (*Agrostis stolonifera*) vegetation. The dry meadow vegetation type exists farther upstream, where water table depths are in the range of 120–180 centimeters. Dry meadow vegetation includes mat muhly (*Muhlenbergia richardsonis*), Nevada bluegrass (*Poa secunda*), and slender wheatgrass *Elymus trachycaulus*). The basin big sagebrush–basin wildrye type is located in the upper portion of the Big Creek site where the depth to groundwater is deepest and subject to the greatest spatial and temporal variations (fig. 5.3). Basin big sagebrush (*Artemisia tridentata tridentata*), basin wildrye (*Leymus cinereus*), and bluegrass (*P. secunda*) comprise this vegetation type. A similar correlation between vegetation and depth to groundwater was observed at other Great Basin riparian meadow study areas, including at Indian Creek valley (Castelli et al. 2000; Lord et al. in review; chapter 7).

Impact of Stream Incision on Hydrology and Vegetation

The entrenchment of much of Indian Creek valley has strongly affected the groundwater and vegetational characteristics of the study site. The depth to the groundwater table is less than 1 meter in unincised wet and mesic meadows and typically several meters deep adjacent to incised trenches (table 5.1). The drop in the groundwater table is apparent from well data (e.g., at transect 10) and by viewing groundwater seeps in trench headcuts, which slant downward in the direction of groundwater flow. Although the main incision at Indian Creek valley resulted in the trench, the trench bottom has been incised a couple of decimeters in places, resulting in one or two terraces. As with other meadow complexes in the central Great Basin, vegetation type is strongly correlated with depth to groundwater (Lord et al. in review; chapter 7). Consequently, the vegetation types at Indian Creek correlate with the geomorphic setting because of its relationship to the depth to groundwater (table 5.2). Vegetation type is most sensitive to the geomorphic setting (i.e., vertical position) in the upstream area where there is less variability in groundwater levels. For example, at transect 7, wet meadow vegetation is present at the lowest level of the trench floor, mesic vegetation is on the trench floor terrace, only 10 to 15 centimeters above the lowest level, and sagebrush is dominant above the trench on the valley floor.

Many incised trenches at Indian Creek valley are actively migrating upgradient (fig. 5.9). Trench sides are steep, and angular slump blocks are common at active seeps. This form is characteristic of erosion caused by groundwater sapping (e.g., Howard and McLane 1988; Kochel et al. 1988; Higgins et al. 1990; Pederson 2001). Once incision is initiated, the process is self-enhancing, commonly causing an increase in erosion rates with time. The incision causes lowering of the groundwater table and an increased convergence of groundwater toward the headcut (Dunne 1990; Montgomery 1999; Pederson 2001). Continued effectiveness of this process requires the removal of failed debris from the slope base by surface waters (Dunne 1990; Higgins et al. 1990). If the failed material is not removed, it may result in discontinuous gullies (Montgomery 1999). The effectiveness of the sapping process is further enhanced because the drop in the water table can cause a decrease or change in vegetation type (Keller et al. 1990). In the case of Indian Creek, the drop has resulted in a shift in vegetation type from wet meadow to dry meadow, mesic meadow, or basin big

TABLE 5.2.

Mean depth to water table (±1 SD) and dominant vegetation types
(TWINSPAN) at different geomorphic positions at an upstream and
downstream transect in Indian Creek valley

Variable		Transect and Well	
Upstream reach well sites			
Well number	7D	7B	7C
Geomorphic setting	Valley floor	Trench floor: upper terrace	Trench floor: lower terrace
Depth to water table	1.65 ± 0.06	0.44 ± 0.23	0.35 ± 0.07
Vegetation type	Artemisia tridentata/ Leymus cinereus	Mesic meadow	Wet meadow
Downstream reach well sites			
Well number	3D	3C	3B
Geomorphic setting	Valley floor	Trench floor: upper terrace	Trench floor: lower terrace
Depth to water table	1.44 ± 0.76	1.25 ± 0.87	1.14 ± 0.88
Vegetation type	Dry meadow and Artemisia tridentata/ Leymus cinereus	Dry meadow	Wet and dry meadow

Vegetation types and analysis follow Weixelman et al. (1996). Modified from Lord et al. in review.

sagebrush meadow (table 5.2, figs. 5.9B and C). In meadow complexes, soils supporting wet meadows have greater bank strength than dry meadow soils and, therefore, are less prone to failure (Micheli and Kirchner 2002a,b). In a study of the influence of bank erosion rates in a meadow complex of California's Sierra Nevada range, Micheli and Kirchner (2002b) found that dry meadow banks retreated six times faster than wet meadow banks. In sum, once entrenchment of a meadow complex initiates, it is highly likely to continue until it reaches a new stable state (Schumm 1999).

The rates of growth of the trenches in Indian Creek valley are not known, but it is clear from a preliminary analysis of historical aerial photographs that they have enlarged significantly over the past several decades. For example, the over 500 meters of trenches circled on the 1993 photo (figs. 5.9A) have developed since 1961. Wet meadow complexes were present on the valley floor downstream of their present location, presumably prior to the groundwater table dropping in response to trench incision. This is indicated by active trenches, the presence of meadow paleosols on the valley floor of entrenched areas, and the abrupt change in vegetation, from wet meadow complexes to sagebrush, at headcut positions. This vertical displacement of broad meadow complexes downward to strip meadows on the trench floor has resulted locally in an approximately 90 percent loss of area in the meadow complexes. As discussed above, it is probable

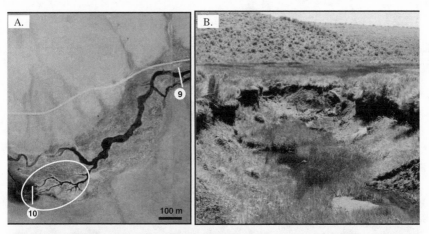

FIGURE 5.9. Actively incising meadow complex: (A) Vertical aerial photo (1993) of upstream portion of study area. The entire trench system circled has developed since 1961. The locations of transects 9 and 10 are marked. (B) Upvalley view of headcut trench (its location is indicated by the arrow marked by a star in (C). The trench is actively migrating upstream as indicated by steep trench sides and an abundance of fresh slump blocks. Wet-meadow vegetation is present on the floor of the trench but not adjacent to the trench on the valley floor because of the drop in the water table caused by incision. (C) Broad meadow complex, in vicinity of transect 10, with three headcut trenches (arrows) at downstream end of meadow. Note the change in vegetation type, from a wet-meadow complex to one dominated by sagebrush and rabbit bush, coincident with the position of the headcut trenches.

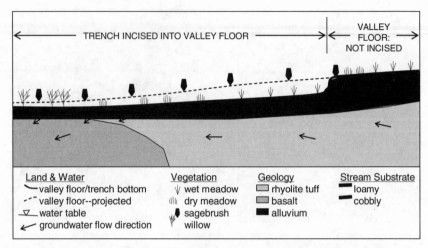

FIGURE 5.10. Schematic cross section through the Indian valley study area showing mutual relationships of bedrock, surface sediment, water table, groundwater flow direction, trench depth, and vegetation.

that the incision of the valley floor will continue in the future as it has in the past. Additionally, the hydrologic and vegetational characteristics that exist downstream will migrate upstream in response to the new incision (fig. 5.10). Eventually, over a long time scale (at least hundreds of years), it is probable that the trench systems will widen to span the valley floor and to once again support broad, wet meadow complexes.

The specific causes that initiated entrenchment of Indian Creek valley are not known at this time. However, the stratigraphy of the Indian Creek valley fill and other work in southwestern United States suggest that episodes of incision of the valley alluvium have occurred in the past and that incision is related to climatic change and basin sediment supply (Chambers et al. 1998; Elliott et al. 1999; chapter 3). Upland axial valleys demonstrate different degrees of sensitivity to incision (Miller et al. 2001; chapter 4). Indian Creek valley is sensitive to entrenchment driven by groundwater sapping. Favorable conditions for sapping present in Indian Creek valley include convergence of groundwater toward a headcut, unconsolidated material, a critical bank height to induce slope failure, and enough surface flow to remove failed debris (Kochel et al. 1988; Dunne 1990; Nash 1996; Pederson 2001). Although natural conditions in Indian Creek valley are conducive to incision, it is likely that land-use activities within the valley have accelerated or triggered some of the entrenchment. Over the past century, and probably before, numerous check dams were constructed to divert water from small valley-floor stream channels, and cattle used the land for grazing. In some places, these activities concentrated runoff to trench headcuts and, therefore, probably enhanced the ability of surface flow to remove failed debris from trench bottoms. In addition, diversion and ponding of water would have altered the elevation of the water table and affected sapping processes.

Management and Restoration Implications

Understanding the geomorphic and hydrogeologic controls on the evolution of riparian habitat in the Great Basin of central Nevada can be used to develop watershed-scale management strategies intended to maintain or restore these ecosystems. The following questions should be addressed in considering management options:

- What are the hydrogeologic conditions that led to the formation of meadow complexes?
- What are the base-level controls on the valley segments and stream reaches of interest? What are the base level controls on the groundwater regime?
- What is the susceptibility of the soils, sediments, and bedrock underlying the meadows to incision by surface water erosion or groundwater sapping?
- What is the sensitivity of the existing vegetation to change with changes in the groundwater regime?

Conceptualizing and characterizing surface water and groundwater interactions at the valley-segment and stream-reach scales is an essential step in restoring riparian ecosystem process and pattern (Woessner 2000; National Research Council 2002). The complexities of the groundwater–surface water interactions in riparian meadow systems in the Great Basin are apparent from the case studies highlighted in this chapter. The hydraulic conductivity, connectivity, and geometry of the alluvial aquifer materials determine the spatial and temporal characteristics of groundwater flow. These geologic and hydrogeologic conditions must be understood so that the appropriate hydrologic regime (both surface and subsurface flows) can be managed or reestablished through restoration activities. Additional studies to investigate smaller-scale stream and groundwater mixing in the hyporheic zone and to evaluate the influence of bedrock structure at these and other riparian meadow study sites are under design.

Determining the base-level controls on stream and groundwater levels is important for maintaining sust/ainable riparian ecosystems and developing restoration strategies for degraded ecosystems. The potential maximum depth of stream entrenchment in upland watersheds must be assessed at sites warranting restoration. Restoration alternatives need to be evaluated with respect to this potential base level so that viable alternatives can be identified. For example, the selection of channel or stream bank stabilization alternatives depends on the likelihood of these stabilization efforts being undermined by erosion associated with future high-intensity rainfall and runoff events. Because groundwater levels will likely adjust themselves to newly established base levels following stream incision, local hydrogeologic and stratigraphic controls also must be considered during the base-level evaluation.

It is necessary to understand the incision processes and sensitivity of a stream system to incision because of the importance of streams in establishing local base levels. As discussed in chapter 4, upland basins of the central Great Basin may be grouped by characteristics that correlate with sensitivity to incision. Riparian ecosystems that are more sensitive require careful management and, if degraded, will require more effort to restore. Furthermore, incision of the valley floor by stream downcutting versus entrenchment driven by groundwater sapping require different management strategies. Stream downcutting progresses by the upstream migration of knickpoints along the stream bed (chapter 4). To slow incision of riparian meadows in these areas, the knickpoints should be stabilized, especially in and immediately downstream from meadow complexes. In meadow complexes prone to or exhibiting signs of groundwater sapping, the headcuts must be stabilized. As described previously, once incision by groundwater sapping is triggered, the process is self-enhancing and will be difficult to stop. Understanding the directions of groundwater flow and stratigraphy is important, because trenches extend preferentially along pathways of higher groundwater discharge (Kochel et al. 1988).

The sensitivity of plant species composition to the groundwater regime is obvious at both the Big Creek and the Indian Creek study areas. Vegetation type and pattern are determined by the hydrologic regime in riparian areas of the semiarid and arid United States (Stromberg et al. 1996; Martin and Chambers 2001; chapter 7). A lowering of the water table following stream incision can reduce the areal extent of meadow and other riparian vegetation types, and significantly alter plant species composition and diversity. In some settings, alternative stable vegetation states can be used to create a mosaic of *Artemisia* and drier meadow ecosystems that does not duplicate but more closely resembles the conditions that existed prior to incision (Chambers and Linnerooth 2001; Wright and Chambers 2002). In order to accurately evaluate the effects of stream incision and the subsequent lowering of water tables on riparian plant species and communities, we must first understand the relationships among vegetation composition and pattern and groundwater regimes.

Stream incision is a major issue affecting the sustainability of riparian ecosystems in upland watersheds in central Nevada. In order to protect threatened riparian ecosystems and to restore degraded riparian ecosystems, one must have a thorough knowledge of the hydrologic regime. In fact, a National Research Council report on riparian area functions and management strategies states that "repairing the hydrology of the system is the most important element of riparian restoration" (National Research Council 2002). Still, the relationship between hydrology, geomorphology, and vegetation patterns and dynamics is complex (Patten 1998; Tabacchi et al. 1998; chapter 7). A successful riparian ecosystem management and restoration strategy requires a thorough and integrated understanding of all of these issues.

Acknowledgments

The authors want to thank Robert Barr, Kurtis Beshers, Chris Bocchichio, Christiana Bruinsma, Ronnie Dilbeck, Chris Ernst, Reed Johnson, John Korfmacher, Joshua Lord, Dorothea Richardson, Rob Vartabedian, Sarah Vartabedian and Karen Waltman for their assistance with data collection efforts. Bob E. Hall and Kathy Tynsky provided valuable support in preparing maps and figures. The manuscript was improved through the helpful reviews provided by Dr. Dru Germanoski, Dr. Greg Pohll and Steve Vandegrift. Funding for this research was provided in part by the USDA Forest Service Stream Systems Technology Center and the Great Basin Ecosystem Management Project. The U.S. Environmental Protection Agency through its Office of Research and Development also partially funded and collaborated in the research described here under its Ecosystem Restoration Program. This chapter has not been subjected to Agency review and therefore does not necessarily reflect the views of the Agency, and no official endorsement should be inferred.

Literature Cited

Allen-Diaz, B. 1991. Water table and plant species relationships in Sierra Nevada meadows. *American Midland Naturalist* 126:30–43.

Bendix, J., and C. R. Hupp. 2000. Hydrological and geomorphological impacts on riparian plant communities. *Hydrological Processes* 14:2977–2990.

Berger, D. L. 2000. *Water budgets for Pine Valley, Carico Lake Valley, and Upper Reese River Valley hydrographic areas, middle Humboldt River Basin, North-central Nevada: Methods for estimation and results.* U.S. Geological Survey, Water-Resources Investigations Report 99-4272. Carson City, NV.

Bull, W. B. 1997. Discontinuous ephemeral channels. *Geomorphology* 19:227–276.

Castelli, R. M., J. C. Chambers, and R. J. Tausch. 2000. Soil-plant relations along a soil-water gradient in Great Basin riparian meadows. *Wetlands* 20:251–266.

Chambers, J. C., K. Farleigh, R. J. Tausch, J. R. Miller, D. Germanoski, D. Martin, and C. Nowak. 1998. Understanding long- and short-term changes in vegetation and geomorphic processes: The key to riparian restoration. Pp. 101–110 in *Rangeland management and water resources*, edited by D. F. Potts. Herdon, Va.: American Water Resources Association and Society for Range Management.

Chambers, J. C., and A. R. Linnerooth. 2001. Restoring riparian meadows currently dominated by *Artemisia* using alternative state concepts — the establishment component. *Applied Vegetation Science* 4:157–166.

Cooper, D. J. 1994. *Sustaining and restoring western wetland and riparian ecosystems threatened by or affected by water development projects.* General Technical Report RM-247, USDA Forest Service, Rocky Mountain Research Station. Fort Collins, Colo.

Dunne, T. 1990. Hydrology, mechanics, and geomorphic implications of erosion by subsurface flow. Pp. 1–28 in *Groundwater geomorphology: The role of subsurface water in earth-surface processes and landforms*, edited by C. G. Higgins and D. R. Coates. Special Paper 252. Geological Society of America. Boulder, Colo.

Elliott, J. G., A. C. Gellis, and S. B. Aby. 1999. Evolution of arroyos: Incised channels of the southwestern United States. Pp. 153–185 in *Incised river channels*, edited by S. E. Darby and A. Simon. Chichester, UK: John Wiley & Sons.

Feller, J. M. 1998. Recent developments in the law affecting livestock grazing on western riparian areas. *Wetlands* 18:646–657.

Fetter, C. W. 2001. *Applied hydrogeology*, 4th ed. Upper Saddle River, N.J.: Prentice-Hall.

Golden Software. 1995. SURFER contouring and 3D surface mapping program and user's guide. Golden, Colo.: Golden Software, Inc.

Harvey, J. W., and K. E. Bencala. 1993. The effect of streambed topography on surface-subsurface water exchange in mountain catchments. *Water Resources Research* 29:89–98.

Heath, R. C. 1988. Hydrogeologic settings of regions. Pp. 15–23 in *Hydrogeology, the geology of North America*, edited by W. Back, J. S. Rosenshein, and P. R. Seaber. Vol. O-2. Boulder, Colo.: Geological Society of America.

Hess, G. W., and L. R. Bohman. 1996. *Techniques for estimating monthly mean streamflow at gaged sites and monthly streamflow duration characteristics at ungaged sites in central Nevada.* Open File Report 96-559. U.S. Geological Survey. Carson City, NV.

Higgins, C. G., B. R. Hill, and A. K. Lehre. 1990. Gully development. Pp. 139–155 in *Groundwater geomorphology: The role of subsurface water in earth-surface processes and landforms*, edited by C. G. Higgins and D. R. Coates. Special Paper 252. Geological Society of America. Boulder, Colo.

Hill, B. R. 1990. Groundwater discharge to a headwater valley, northwestern Nevada, USA. *Journal of Hydrology* 113:265–283.

Howard, A. D., and C. F. McLane III. 1988. Erosion of cohesionless sediment by groundwater seepage. *Water Resources Research* 24:1659–1674.

Hubbard, J. P. 1977. Importance of riparian ecosystems: Biotic considerations. Pp. 14–18 in *Importance, preservation, and management of riparian habitat: A symposium*, edited by R. R. Johnson and D. A. Jones. General Technical Report RM-43, USDA Forest Service. Fort Collins, Colo.

Johnson, R. R., and S. W. Carothers. 1982. Riparian habitats and recreation: Interrelationships in the Southwest and Rocky Mountain region. USDA Forest Service, Rocky Mountain Forest and Range Experiment Station, Fort Collins, Colo. *Eisenhower Consortium Bulletin* 12:1–31.

Jones, J. B., and P. J. Mullholland. 2000. *Streams and ground water*. San Diego: Academic Press.

Kaufman, J. B., and W. C. Krueger. 1984. Livestock impacts on riparian ecosystems and streamside management implications . . . a review. *Journal of Range Management* 37:430–438.

Keller, E. A., G. M. Kondoff, D. J. Hagery, and G. M. Kondoff. 1990. Groundwater and fluvial processes: Selected observations. Pp. 319–340 in *Groundwater geomorphology: The role of subsurface water in earth-surface processes and landforms*, edited by C. G. Higgins and D. R. Coates. Special Paper 252. Geological Society of America. Boulder, Colo.

Kleinhampl, F. J., and J. I. Ziony. 1985. *Geology of northern Nye County, Nevada*. Bulletin 99A. Nevada Bureau of Mines and Geology. University of Nevada, Reno.

Kochel, R. C., D. W. Simmons, and J. F. Piper. 1988. Groundwater sapping experiments in weakly consolidated layered sediments: A qualitative summary. Pp. 84–93 in *Sapping features of the Colorado Plateau: A comparative planetary geology field guide*, edited by A. D. Howard, R. C. Kochel, and H. E. Holt. NASA SP-491. Washington, D.C.: National Aeronautics and Space Administration.

Law, D. J., C. B. Marlow, J. C. Mosley, S. Custer, P. Hook, and B. Leinard. 2000. Water table dynamics and soil texture of three riparian plant communities. *Northwest Science* 74:233–241.

Lord, M. L., J. R. Miller, and J. C. Chambers. *In review*. Geomorphic, hydrologic, and vegetational response of meadow complexes to valley incision in the central Great Basin. *Wetlands*.

Martin, D. W., and J. C. Chambers. 2001. Effects of water table, clipping, and species interactions on *Carex nebrascensis* and *Poa pratensis* in riparian meadows. *Wetlands* 21:422–430.

———. 2002. Restoration of riparian meadows degraded by livestock grazing: Above- and below-ground responses. *Plant Ecology* 163:77–91.

Maxey, G. B. 1968. Hydrogeology of desert basins. *Ground Water* 6:10–22.

Maxey, G. B., and T. E. Eakin. 1949. *Ground water in the White River valley, White Pine, Nye, and Lincoln counties, Nevada*. Nevada State Engineer, Water Resources Bulletin 8.

McKee, E. H. 1976. *Geologic Map of the Austin Quadrangle, Lander County, Nevada*. Nevada Bureau of Mines and Geology, Map GQ-1307, 1:62,500 scale.

Micheli, E. R., and J. W. Kirchner. 2002a. Effects of wet meadow riparian vegetation on stream bank erosion. 1. Remote sensing measurements of streambank migration and erodibility. *Earth Surface Processes and Landforms* 27:627–639.

——. 2002b. Effects of wet meadow riparian vegetation on stream bank erosion. 2. Measurements of vegetated bank strength and consequences for failure mechanics. *Earth Surface Processes and Landforms* 27:687–697.

Mifflin, M. D. 1988. Region 5, Great Basin. Pp. 69–78 in *Hydrogeology, the geology of North America*, edited by W. Back, J. S. Rosenshein, and P. R. Seaber. Vol. O-2. Boulder, Colo.: Geological Society of America.

Miller, J. R., D. Germanoski, K. Waltman, R. J. Tausch, and J. C. Chambers. 2001. Influence of late Holocene hillslope processes and landforms on modern channel dynamics in upland watersheds of central Nevada. *Geomorphology* 38:373–391.

Montgomery, D. R. 1999. Erosional processes at an abrupt channel head: Implications for channel entrenchment and discontinuous gully formation. Pp. 247–276 in *Incised river channels*, edited by S. E. Darby and A. Simon. Chichester, UK: John Wiley & Sons.

Naiman, R. J., and H. Decamps. 1997. The ecology of interfaces: Riparian zones. *Annual Review of Ecology and Systematics* 28:631–658.

Nash, D. J. 1996. Groundwater sapping and valley development in the Hackness Hills, North Yorkshire, England. *Earth Surface Processes and Landforms* 21:781–795.

National Research Council. 2002. *Riparian areas: Functions and strategies for management*. Washington, D.C.: National Academy Press.

Nichols, W. D. 2000. *Regional groundwater evapotranspiration and groundwater budgets, Great Basin, Nevada*. Professional Paper 1628. Reston, VA: U.S. Geological Survey.

Ohmart, R. D., B. W. Anderson, and W. C. Hunter. 1988. The ecology of the lower Colorado River from Davis Dam to the Mexico–United States boundary: A community profile. *U.S. Fish and Wildlife Biological Report* 85:1–296.

Patten, D. T. 1998. Riparian ecosystems of semi-arid North America: Diversity and human impacts. *Wetlands* 18:498–512.

Pederson, D. T. 2001. Stream piracy revisited. *GSA Today* 11:4–10.

Saab, V., and C. Groves. 1992. *Idaho's migratory birds: Description, habitats, and conservation*. Nongame Wildlife Leaflet 10. Boise, Idaho: Department of Fish and Game.

Schumm, S. A. 1999. Causes and controls on channel incision, in incised river channels. Pp. 19–33 in *Incised river channels*, edited by S. E. Darby and A. Simon. Chichester, UK: John Wiley & Sons.

Stanford, J. A., and J. V. Ward. 1988. The hyporheic habitat of river ecosystems. *Nature* 335:64–66.

Stewart, J. H., and J. E. Carlson. 1976. *Geologic map of north-central Nevada*. Nevada Bureau of Mines and Geology, Map 50, 1:250,000 scale. Reno, NV.

Stromberg, J. C., R. Tiller, and B. Richter. 1996. Effects of groundwater decline on riparian vegetation of semi-arid regions: The San Pedro, Arizona. *Ecological Applications* 6:113–131.

Tabacchi, E., D. L. Correll, G. Pinay, A. M. Planty-Tabacchi, and R. C. Wissmar. 1998. Development, maintenance and role of riparian vegetation in the river landscape. *Freshwater Biology* 40:497–516.

Valett, H. M., S. Fisher, and E. H. Stanley. 1990. Physical and chemical characteristics of the hyporheic zone of a Sonoran Desert stream. *Journal of the North American Benthological Society* 9:201–205.

Weixelman, D. A., D. C. Zamudio, and K. A. Zamudio. 1996. *Central Nevada riparian field guide*. R4-ECOL-TP. USDA Forest Service, Intermountain Region. Ogden, Utah.

Winter, T. C. 1999. Relation of streams, lakes, and wetlands to groundwater flow systems. *Hydrogeology Journal* 7:28–45.

Winter, T. C., J. W. Harvey, O. L. Franke, and W. M. Alley. 1998. *Ground water and surface water: A single resource*. Circular 1139. U.S. Geological Survey. Denver, Colo.

Woessner, W. W. 1998. Changing views of stream-groundwater interaction. Pp. 1–6 in *Proceedings of the Joint Meeting of the 28th Congress of the International Association of Hydrogeologists and the annual meeting of the American Institute of Hydrology*, edited by J. Van Brahana, Y. Eckstein, L. W. Ongley, R. Schneider, and J. E. Moore. American Institute of Hydrology. St. Paul, Minn.

———. 2000. Stream and fluvial plain ground water interactions: Rescaling hydrogeologic thought. *Ground Water* 38:423–429.

Wright, J. M., and J. C. Chambers. 2002. Restoring riparian meadows currently dominated by *Artemisia* using alternative state concepts. *Applied Vegetation Science* 5:237–246.

Wroblicky, G. J., M. E. Campana, H. M. Valett, and C. N. Dahm. 1998. Seasonal variation in surface-subsurface water exchange and lateral hyporheic area of two stream-aquifer systems. *Water Resources Research* 34:317–328.

Chapter 6

Effects of Natural and Anthropogenic Disturbances on Water Quality

Michael C. Amacher, Janice Kotuby-Amacher, and Paul R. Grossl

Most of the land area within the Monitor, Toquima, and Toiyabe Ranges of central Nevada falls within the boundaries of the Humboldt-Toiyabe National Forest and is managed by the USDA Forest Service. Past and present land-use activities consist primarily of mineral exploration, mining, domestic livestock grazing, and recreational activities such as road and off-road travel, camping, hunting, and fishing. Numerous intermittent and perennial streams are present in these mountain ranges, yet little is known about the impact of these land-use activities on stream water quality. Aside from land use, water quality in natural forest and range ecosystems is influenced by many factors including atmospheric deposition, catchment lithology, vegetation, stream discharge (controlled by episodic climatic events), natural disturbances (e.g., wildfires), and land management activities (e.g., prescribed burning for fuel reduction and plant community conversion).

Because of the critical importance of the more abundant water resources on national forest lands to overall ecosystem health and economic development of local communities in comparison to the much drier desert basins, an assessment of the factors influencing water quality of streams in these mountain ranges was conducted.

Methods

A broadscale water quality survey was conducted to assess overall water quality in most of the named streams in the Monitor, Toquima, and Toiyabe Ranges in central Nevada. Subsequent to this analysis, additional site- and seasonal-intensive sampling of selected streams in the Toiyabe Range was conducted to assess within catchment spatial and seasonal water quality variability.

162

1994 Water Quality Survey

Water samples for this survey were collected in April and May of 1994. Streams to be sampled were selected from the 1987 Toiyabe National Forest travel maps for the Austin and Tonopah Ranger Districts (USDA Forest Service, Intermountain Region, Ogden, Utah) based on catchment size and road access. To ensure a large sample size, virtually the entire population of named streams in these mountain ranges that had road access to their canyons was selected for sampling. This population included perennial, seasonal, and intermittent streams. Although attempts were made to visit all the named streams, access problems in some canyons and dry channels prevented a complete sampling. Snowpack during the winter of 1994 was low. Thus, many streams were dry at the time they were visited. A total of thirty-nine catchments were sampled.

Sampled streams and catchment lithology are summarized in table 6.1. Approximate locations are indicated on a map of the area (fig. 6.1). Because the number of canyons to be visited was large and covered thousands of square miles of areal extent, sampling locations were selected primarily based on access and proximity to the mouth of the catchments. Thus, water samples collected at the base of the catchments reflect an integrated assessment of water quality delivered to that point from the entire catchment on national forest lands.

Stream water temperature and pH were measured on site using an Orion SA250 portable pH meter, a Ross combination pH electrode, and

TABLE 6.1.

Lithologies of catchments in the Monitor, Toquima, and Toiyabe Ranges of central Nevada

Lithology	Mountain Range	Catchment Streams
Igneous	Monitor	Faulkner, Allison, Willow (south), Barley, Mosquito, Morgan, Willow (west), Reynolds
	Toquima	Pine, Andrews, Willow, Barker
	Toiyabe	Birch, Reese, Stewart, Clear, Tierney
Igneous/sedimentary	Monitor	Clear, Danville
	Toquima	Stoneberger, Meadow, Shoshone, Jefferson
	Toiyabe	Alice Gendron, Ophir, Last Chance, North Twin, South Twin, Jett, Marysville, Cottonwood, San Juan, Washington
Sedimentary	Monitor	None
	Toquima	None
	Toiyabe	Lynch, Tar, Kingston, Bowman, Aiken, Summit, Big

Roberts et al. 1967; Stewart and McKee 1977; Kleinhampl and Ziony 1985

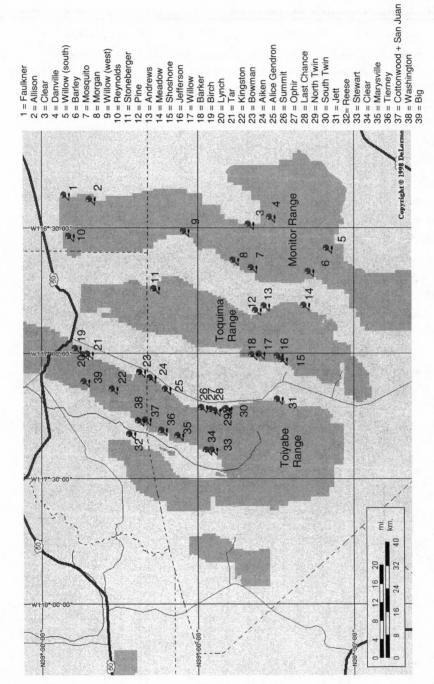

1 = Faulkner
2 = Allison
3 = Clear
4 = Danville
5 = Willow (south)
6 = Barley
7 = Mosquito
8 = Morgan
9 = Willow (west)
10 = Reynolds
11 = Stoneberger
12 = Pine
13 = Andrews
14 = Meadow
15 = Shoshone
16 = Jefferson
17 = Willow
18 = Barker
19 = Birch
20 = Lynch
21 = Tar
22 = Kingston
23 = Bowman
24 = Aiken
25 = Alice Gendron
26 = Summit
27 = Ophir
28 = Last Chance
29 = North Twin
30 = South Twin
31 = Jett
32= Reese
33 = Stewart
34 = Clear
35 = Marysville
36 = Tierney
37 = Cottonwood + San Juan
38 = Washington
39 = Big

Copyright © 1998 DeLorme

FIGURE 6.1. Monitor, Toquima, and Toiyabe Ranges of central Nevada, and approximate stream sampling locations.

an automatic temperature compensator. The meter and electrodes were calibrated with pH 4.00 and 7.00 buffers immediately prior to each pH and temperature measurement. Stream discharge (flow) was determined by measuring cross-sectional area (distance across stream × depth) and flow velocity using a Marsh-McBirney model 201D portable water current meter (Gordon et al. 1992). At each sampling site, a grab-sample of stream water was collected in a 500-milliliter widemouthed, plastic bottle. The bottles were cleaned with phosphorus-free detergent, soaked in a dilute nitric acid bath, and rinsed several times with deionized water prior to use. Water samples were stored in coolers until returned to the lab.

Upon return to the lab, all water samples were filtered through 0.4-micromillimeter polycarbonate membrane filters. A 250-milliliter subsample of each was acidified with 1.25 milliliters of select-grade concentrated HNO_3 (nitric acid). The remaining unacidified samples were stored at less than 4°C and analyzed within two weeks of collection.

Filtered and unacidified water samples were analyzed for alkalinity by the automated methyl orange method, $NO_2 + NO_3$-N by the automated Cd reduction method, and Cl by the automated mercuric thiocyanate method (LACHAT 1991a,b, 1992). Fluoride was determined by the SPADNS colorimetric method and dissolved orthophosphate (o-PO_4) by the ascorbic acid colorimetric method (APHA et al. 1992). Filtered and acidified water samples were analyzed for Na (sodium), K (potassium), Mg (magnesium), Ca (calcium), Sr (strontium), B (boron), Al (aluminum), Si (silicon), and S (sulfur) by inductively coupled plasma atomic emission spectrometry (ICP-AES) (APHA et al. 1992). Trace elements (Cr (chromium), Mo (molybdenum), Mn (manganese), Fe (iron), Co (cobalt), Ni (nickel), Cu (copper), Zn (zinc), Cd (cadmium), Pb (lead), As (arsenic), and Se (selenium)) were determined by extracting a 200-milliliter subsample of filtered, acidified water with 5 percent ammonium pyrrolidine dithiocarbamate (APDC) in chloroform, evaporating the chloroform, redissolving the metals in 50 percent hydrochloric acid, and diluting to a 10-milliliter final volume to concentrate the metals twenty to one prior to analysis by ICP-AES (APHA et al. 1992; Bradford and Bakhtar 1991; U.S. EPA 1976).

1995–1998 Seasonal Water Quality Survey

A site- and seasonal-intensive water quality survey of selected streams in the Toiyabe Range was conducted from 1995 through 1998. Streams sampled included Birch (eight sites), Kingston (five main channel sites, six spring-fed tributaries), Big (ten sites), Washington (six sites), Cottonwood (two sites), San Juan (nine sites), Marysville (four sites), Clear (two sites), and Stewart Creeks (two sites), and the Reese River (three sites). To determine the effect of seasonal flow on water quality parameters, streams were sampled in April or May and in June, July, and September, although not every site was sampled on each visit.

Additional on-site measurements and sample types were added to the intensive water quality survey, including water level and temperature (Aquarod water level loggers), turbidity (measured on-site with a portable turbidity meter using infrared 180° backscatter technology), suspended sediment (sampled with a U.S. Geological Survey DH-81 suspended-sediment sampler by the depth-integrated equal-transit-rate–equal-width-increment method), and bedload (sampled with a 3:22 Helley-Smith bedload sampler using the single equal-width-increment method) (Gordon et al. 1992; Williams et al. 1988).

The DH-81 suspended-sediment sampler was also used to collect water samples for subsequent filtration to determine dissolved constituents. Upon return to the lab, these samples were filtered through 50-millimeter filter units containing 1.0-micrometer glass fiber prefilters and 0.45-micrometer nylon membrane filters using 50-milliliter all-plastic syringes. A 10-milliliter subsample of each filtered sample was acidified with 100 microliters of concentrated nitric acid prior to ICP-AES analysis.

The filtered samples were analyzed for the same dissolved constituents as were determined in the 1994 water quality survey. Additional analyses of unacidified water samples included NH_4-N (LACHAT 1993) and total P by alkaline persulfate digestion followed by colorimetric measurement of o-PO_4 using the ascorbic acid method (Eaton et al. 1995).

Suspended sediment concentrations in unfiltered water samples were determined gravimetrically by subsampling the rapidly stirred sample using a wide-bore pipette. The pipetted subsamples were vacuum filtered through preweighed 2.4-centimeter GF/C glass fiber filter disks in 25-milliliter ceramic Gooch crucibles. The crucibles were dried at 105°C, cooled in a desiccator, and reweighed. To determine the organic matter content of the suspended sediment by loss on ignition, the crucibles were heated in a muffle oven for fifteen minutes at 550°C, cooled in a desiccator, and reweighed.

Hydrogeochemical Modeling

The geochemical modeling program, AquaChem (Waterloo Hydrogeologic, Waterloo, Ontario, Canada), was used to classify the waters based on their geochemical signatures using Piper plots (described in the next section). The chemical composition of the stream waters in equilibrium with several common mineral phases (e.g., amorphous silica, quartz, calcite, gypsum, and hydroxyapatite) were computed to determine if the stream waters were undersaturated, in equilibrium with, or oversaturated with respect to these mineral phases. A charge balance between cations and anions was done as a check for completeness of analysis.

Statistical Analysis

A nonlinear regression procedure was used to determine the relationship, if any, between measured water quality parameters and stream discharge. A one-way analysis of variance was used to test for statistically significant differences among catchment lithologies and locations within a catchment. To account for possible effects of stream discharge on water quality parameters while testing for location differences, stream discharge was used as a covariate. The Tukey test was used for mean comparisons for normally distributed data. If the data were non-normally distributed, then a Kruskal-Wallis one-way analysis of variance on ranks with median comparisons by Dunn's method was used.

Results and Discussion

To assess the effects of land-use activities on water quality, the effects of natural influences such as catchment lithology and stream discharge must first be considered.

Effect of Catchment Lithology on Water Quality

Piper plots are a convenient method to graphically determine unique geochemical signatures of natural surface and groundwaters. Major cation and anion concentrations are plotted on trilinear plots as relative percentages (fig. 6.2). The cation and anion composition points are projected onto the diamond plot above the trilinear plots. Waters with similar geochemical composition or signatures are clustered together at the same location on the diamond plot, whereas waters with different signatures will appear at separate locations on the diamond plot.

In the Monitor Range, the geochemical signatures of Clear, Danville, and Willow (south) Creeks are distinctly different from those of the other streams closely clustered together on the Piper plot (fig. 6.2). In the Toquima Range, Stoneberger Creek's geochemical signature is the most

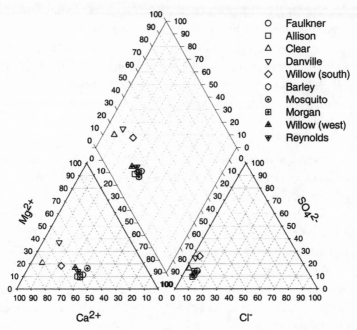

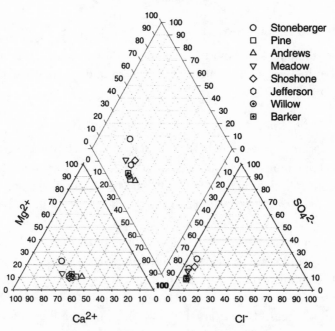

FIGURE 6.2. Piper plots of the chemical composition of streams in the Monitor, Toquima, and Toiyabe Ranges of central Nevada.

Toiyabe Range Streams

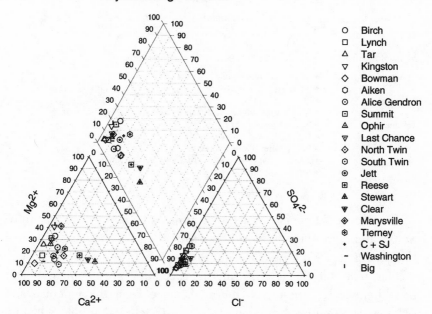

○	Birch
□	Lynch
△	Tar
▽	Kingston
◇	Bowman
○	Aiken
◉	Alice Gendron
▣	Summit
▲	Ophir
▼	Last Chance
◈	North Twin
◉	South Twin
◉	Jett
▣	Reese
▲	Stewart
▼	Clear
◈	Marysville
◉	Tierney
•	C + SJ
-	Washington
ı	Big

different from those of the other streams (fig. 6.2). Meadow, Shoshone, and Jefferson Creeks have similar geochemical signatures and the remaining streams are clustered together in a separate group on the Piper plot. There are several distinct groupings of geochemical signatures of streams in the Toiyabe Range (fig. 6.2). The Reese River along with Stewart and Clear Creeks have the most distinct geochemical signatures. Groups of streams with similar geochemical signatures are listed in table 6.2. Streams with similar geochemical signatures tend to be clustered together geographically and have similar dominant lithologies in their catchments.

The distinct geochemical signatures of some of the streams in the Monitor, Toquima, and Toiyabe Ranges suggest that there may be statistically significant differences in the chemical composition of stream water related to catchment lithology. We classified the catchments into three broad categories: igneous, mixed igneous/sedimentary, and sedimentary based on major rock units found within the catchments (Roberts et al. 1967; Stewart and McKee 1977; Kleinhampl and Ziony 1985). In catchments dominated by mixed igneous and sedimentary rock units, the intense heat of the igneous intrusions altered some of the sedimentary rock units to produce metamorphic rocks. Because of these alterations, metamorphic rocks might not exert a comparable influence on the chemical composition of streams as the original sedimentary rocks from which they were derived.

TABLE 6.2.

Groups of stream with similar geochemical signatures as shown by the Piper plot analysis in Figure 6.2

Mountain Range	Streams with Similar Geochemical Signatures
Monitor	• Clear, Danville, Willow (south)
	• Faulkner, Allison, Barley, Mosquito, Morgan, Willow (west), Reynolds
Toquima	• Stoneberger
	• Meadow, Shoshone, Jefferson, Pine, Andrews, Willow, Barker
Toiyabe	• Birch, Kingston, Summit, Big
	• Lynch, Tar, Bowman, Ophir, Last Chance, Marysville, Washington
	• Jett, Tierney, Cottonwood, San Juan
	• Aiken, Alice Gendron, North Twin, South Twin
	• Reese, Stewart, Clear

Significant differences were found in median Na, Mg, Ca, alkalinity, Si, Cl, and P concentrations of streams draining catchments dominated by igneous rock units as compared to catchments dominated by sedimentary rock units (fig. 6.3). Streams in mixed-lithology catchment streams had chemical compositions intermediate between the igneous- and sedimentary-dominated catchment streams. Catchment lithology had no significant effect on stream discharge, stream water temperature, pH, and K, S, F, and NO_3-N concentrations (fig. 6.3). Even though a statistically significant difference in stream discharge due to catchment lithology was not found in the 1994 water quality survey, catchment lithology can influence stream hydrographs as evidenced by the differences in the Kingston and Big Creek hydrographs (see chapter 4). The 1994 survey was a snapshot of conditions at the time of measurement and did not capture the entire hydrograph.

Further evidence for the effects of catchment lithology on water quality can be found by examining the chemical composition of stream water in catchments where there is a major change in catchment lithology from one part of the catchment to another. The west fork of San Juan Creek flows entirely through Tertiary- and Permian-age volcanic rocks containing feldspar minerals, whereas the east fork borders older Ordovician-, Cambrian-, and Precambrian-age shales, limestones containing carbonate minerals, and locally metamorphosed sedimentary units (Kleinhampl and Ziony 1985). Thus, it is not surprising that the west fork contains significantly higher concentrations of Na, K, and Si than the east fork, while the east fork contains significantly higher concentrations of Mg, Ca, and alkalinity (fig. 6.4). Concentrations of nutrient elements such as S, N, and P in San Juan Creek were not expected to be influenced by the rock formations found in that catchment since their source is largely biological. Thus, the concentrations of these elements in the two forks of San Juan Creek are similar (fig. 6.4).

The highest Si concentrations were found in streams draining catchments dominated by igneous rocks, whereas the lowest concentrations were found in sedimentary rock-dominated catchments (fig. 6.5a) reflecting the greater abundance of amorphous silica in catchments with igneous intrusions. All the streams in the Monitor, Toquima, and Toiyabe Ranges are supersaturated with respect to quartz and are unsaturated with respect to amorphous silica (fig. 6.5a). Because these are surface waters, dilution by snowmelt and shallow soil water could account for undersaturation with respect to amorphous Si when groundwaters that were possibly in equilibrium with amorphous Si in igneous rock units mixed with surface waters.

Stream water alkalinity (as milligrams per liter of $CaCO_3$) generally increased with stream water pH (fig. 6.5b). The highest alkalinity levels tended to be found in catchments dominated by sedimentary rock units such as limestone and dolomite. Calcite is the principal carbonate mineral found in limestone rock units in the mountain ranges of central Nevada. Calcite saturation indexes of the streams increased with increasing pH (fig. 6.5b). High pH streams tended to be supersaturated with respect to calcite, whereas the lower pH streams tended to be undersaturated. Streams fed by groundwater seeps and springs in equilibrium with calcite tend to lose CO_2 by outgassing when the groundwaters contact surface air. As CO_2 is lost from the stream water, pH increases, and the streams become supersaturated with respect to calcite. Whether calcite precipitation from surface stream waters actually occurs depends on reaction kinetics, which is subject to many constraints. For example, the presence of dissolved organic carbon and other constituents in surface and groundwaters can inhibit calcite precipitation even if the waters are supersaturated (Inskeep and Bloom 1986; Doner and Grossl 2002).

A plot of bicarbonate ion concentration against Ca concentration in the stream waters shows that the points tend to fall along the $2[Ca^{2+}] = [HCO_3^-]$ line, where the brackets refer to bicarbonate and Ca ion concentrations in the stream waters (fig. 6.5c). The equation for reaction of calcite with CO_2 and water is

$$CaCO_3 + CO_2 + H_2O = Ca^{2+} + 2HCO_3^-$$

Calcite equilibrium lines for different atmospheric CO_2 concentrations (expressed as the negative log of the CO_2 pressure in atmospheres) are also

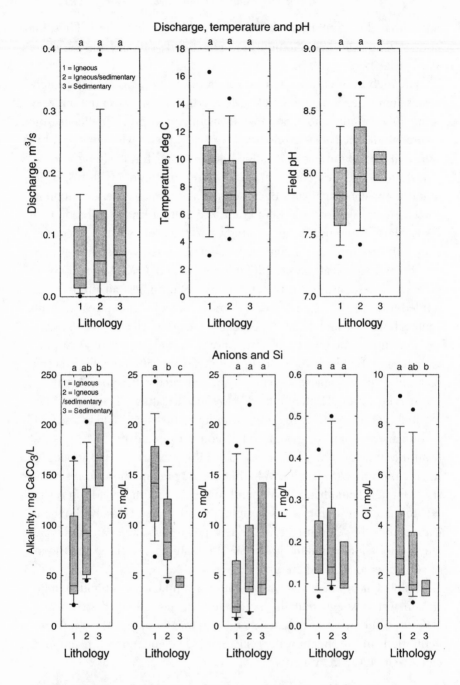

Figure 6.3. Box plots of water quality parameters in the Monitor, Toquima, and Toiyabe Ranges stratified by major catchment lithology (1 = igneous rocks, 2 = igneous and sedimentary rocks, and 3 = sedimentary rocks). The 25th and 75th percentiles are shown as a box around the 50th percentile, the 10th and 90th

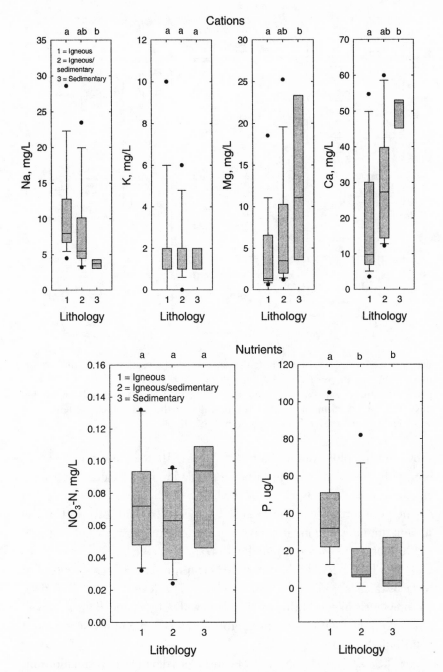

percentiles are shown as error bars, and the 5th and 95th percentiles are shown as points. Water quality parameter means for each catchment lithology not indicated by the same lowercase letter at the top of each plot are significantly different (p < 0.01).

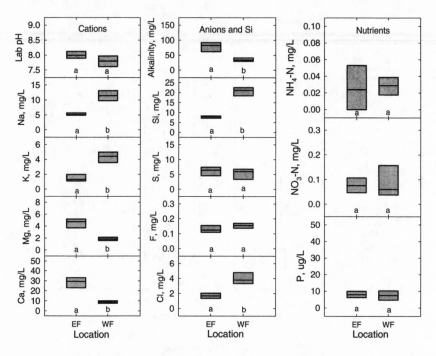

FIGURE 6.4. Box plots of water quality parameters at several locations in the San Juan Creek catchment (EF = East Fork, WF = West Fork). The 25th and 75th percentiles are shown as a box around the 50th percentile. Water-quality parameter means for each catchment location not indicated by the same lowercase letter above each x-axis are significantly different (p<0.01).

shown on the plot. Ambient air CO_2 level is 10–3.5 atmosphere. Thus, most of the streams draining carbonate rock-dominated catchments have alkalinity and Ca levels in excess of what would be observed by assuming calcite equilibrium with ambient CO_2 levels. In contrast, most of the streams draining igneous and mixed igneous/sedimentary–rock-dominated catchments were undersaturated with respect to calcite. Higher levels of CO_2 may be expected in groundwaters (Stumm and Morgan 1996), but surface waters are in contact with ambient CO_2 levels. Thus, groundwaters previously in equilibrium with calcite will outgas some of the dissolved CO_2 on contact with air as these groundwaters come to the surface and contribute to stream flow.

Unlike alkalinity, dissolved P concentrations did not show a relationship with stream water pH (fig. 6.5d). The higher P concentrations tended to be associated with catchments in igneous rock units. Even though dissolved P did not vary with pH, hydroxyapatite saturation indexes tended to increase with stream water pH indicating that the higher pH streams were supersaturated with respect to hydroxyapatite. The equation for precipitation/dissolution of hydroxyapatite in water is

$$Ca_5 (PO_4)_3OH + 7H^+ = 5Ca^{2+} + 3H_2PO_4^- + H_2O$$

Calcium phosphate minerals tend to be the dominant form of P in calcareous soils. Hydroxyapatite is the most thermodynamically stable member of the apatite group, but figure 6.5d shows that it is not controlling dissolved P concentrations in central Nevada streams. Precipitation kinetics have an important role in the formation of Ca phosphate minerals because dissolved organic acids can inhibit mineral crystal formation and growth (Grossl and Inskeep 1992). Since P is a nutrient element, uptake and release of P from terrestrial and aquatic microorganisms, plants, and animals are the likely controlling processes responsible for variations in dissolved P concentrations in the streams of central Nevada.

Effect of Stream Discharge on Water Quality

Although catchment lithology is largely responsible for the basic chemical composition of stream waters in the mountain ranges of central Nevada, other environmental factors can influence the concentrations of constituents in stream waters. One of the most important of these is discharge. Concentrations may increase, remain the same, or decrease with seasonal changes in discharge. Stream discharge in the Monitor, Toquima, and Toiyabe Ranges is highly seasonal (e.g., fig. 6.6). Peak flows for most streams occur in June and are a combination of snowmelt and spring storms, which add additional snowpack and rain to the catchments.

The relationships between discharge and several water quality parameters in Birch Creek in the Toiyabe Mountains are shown in figure 6.7A. These relationships are fairly typical for most streams in the mountain ranges of central Nevada. Generally, concentrations of major constituents such as Mg, Ca, alkalinity, and S in streams with higher concentrations of these constituents such as Birch, Kingston, Big, Cottonwood, San Juan, Marysville, and Reese decreased as discharge increased as a result of snowmelt dilution of stream water (fig. 6.7A). Only Washington, Clear, and Stewart Creeks did not show a significant relationship between discharge and major constituent concentrations.

At very high flows, major constituent concentrations increased in streams with higher concentrations of these constituents producing a quad-

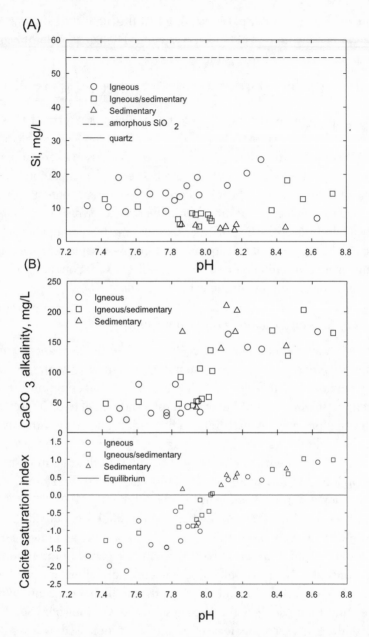

Figure 6.5. (A) Stream water silica (as Si) concentrations stratified by major catchment lithology as a function of pH. The solubility line of quartz is shown as a solid line and the solubility line of amorphous SiO_2 is shown as a dashed line. Si concentrations below a given solubility line are undersaturated with respect to that mineral. (B) Stream water alkalinity (as milligrams per liter of $CaCO_3$) values and calcite saturation indexes stratified by major catchment lithology as a function of pH. A calcite saturation index of 0 indicates that the stream water is in

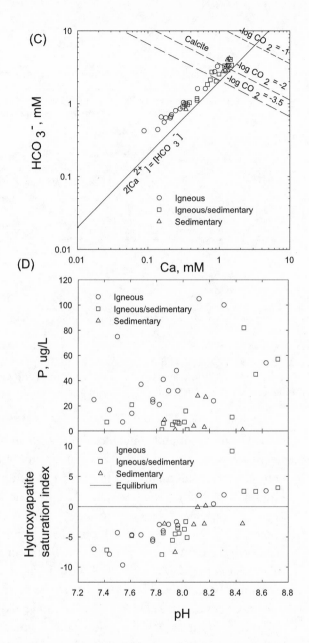

equilibrium with calcite. (C) Charge balance plot of HCO₃⁻ concentrations versus Ca²⁺ concentrations stratified by major catchment lithology (top right). Calcite equilibrium lines at various CO_2 concentrations are shown as dashed lines. (D) Stream water dissolved o-PO₄ (as P) concentrations and hydroxyapatite saturation indexes stratified by major catchment lithology as a function of pH. A hydroxyapatite saturation index of 0 indicates that the stream water is in equilibrium with hydroxyapatite.

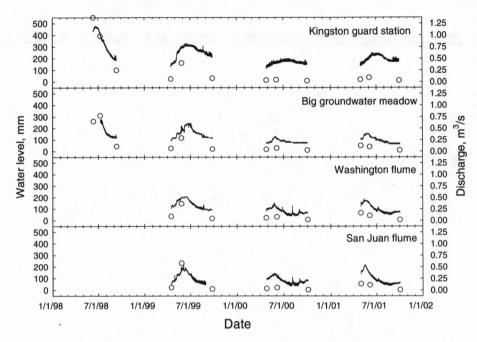

FIGURE 6.6. Seasonal water level and discharge of Kingston, Big, Washington, and San Juan Creeks (1998–2001). Water levels are shown as lines and are associated with the left axis. Discharge values are shown as open circles and are associated with the right axis.

ratic relationship between constituent concentration and stream flow (fig. 6.7A). This concentration increase at high flows is the result of flushing solutes from catchment soils under saturated conditions. Because precipitation in the central Great Basin is so seasonal and highly variable from month to month and year to year, solutes tend to accumulate in catchment soils during dry years and are flushed from the catchments during extreme wet years. The data in figure 6.7A cover a period from 1994 through 1998. Two of the years in that period (1995 and 1998) were high water years, when landscape-flushing bankfull and flood flows were observed.

Generally, concentrations of Na, K, F, Cl, Si, NH_4-N, and P did not vary significantly with discharge. One notable exception is that P concentrations increased with stream flow in Birch Creek (fig. 6.7A), but this stream has higher P concentrations than the other streams in the Toiyabe Range. Nitrate concentrations generally increased with increasing discharge (e.g., fig. 6.7a), which fits the hypothesis of flushing solutes from the catchments at high flows because NO_3-N is very mobile and would be expected to be leached from soil profiles in wet years and accumulate in soils during dry periods.

Although discharge influences the chemical composition of stream water, it probably exerts a larger influence on sediment transport (fig. 6.7B). Suspended sediment and bedload discharge rates increased linearly with stream flow when plotted on a log scale (fig. 6.7B). In general, bedload discharge was greater than suspended sediment discharge during the high water year of 1995 when these data were collected (fig. 6.7B). That sediment transport should be high during high water years is to be expected because it is during these events that most channel incision occurs (see chapters 3 and 4). The erosion and sediment transport and redeposition that occurs during these episodic high-water events may alter the geomorphology and ecology of the streams but appear to have little lasting effect on water quality.

Effect of Land-Use Activities on Water Quality

Land-use activities on national forest lands in the Monitor, Toquima, and Toiyabe Ranges consist largely of cattle and sheep grazing and recreational activities such as vehicle, horseback, and foot travel, and hunting, fishing, and camping. On Bureau of Land Management lands and on private lands in the desert valleys between the mountain ranges, livestock grazing and some irrigated crop production are found. Crop production consists mostly of irrigated alfalfa and grass hay.

Dissolved nutrient concentrations, suspended sediment concentrations, and amounts of bedload sediment are the water quality parameters most likely to be influenced by land-use activities. Waste from grazing animals, improperly disposed-of human wastes, other campground wastes (e.g., detergents), and wastes from cleaning fish and game animals may all contribute to nutrient loads in streams. Land-use activities such as grazing, roads, trails, campsites, and off-road travel, or any activity that degrades soil cover and promotes soil erosion may contribute to increased sediment transport in streams.

Because all these activities are widespread throughout the canyons in the Monitor, Toquima, and Toiyabe Ranges, no obvious land-use contributions to increased sediment transport could be discerned—catchments with the highest suspended sediment concentrations and bedload transport rates had no greater intensity of land-use activities than less-intensively used catchments. Streambank soil erosion rates are related to stream flow, gradient, and geomorphology; soil erodibility; and vegetation root density. Erodibility is in turn influenced by the physical and chemical properties

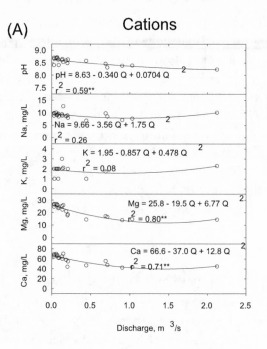

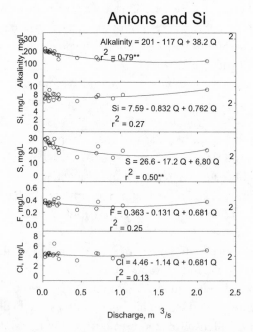

Figure 6.7. (A) Quadratic relationships between water quality parameters and stream discharge for Birch Creek. Double asterisks (**) after the r^2 values indicate that the regression is significant at the 0.01 level of probability. (B, next page) Relationship between suspended and bedload sediment discharge and stream discharge for several streams in the Toiyabe Range in 1995.

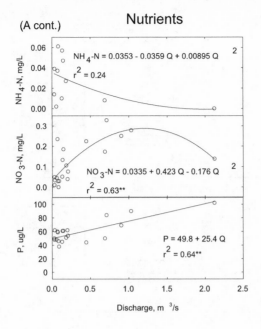

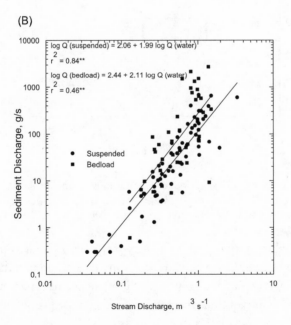

of streambank soils. Coarse-textured soils without cementing agents such as carbonates were found to be more erodible than finer-textured soils that contained carbonates (Amacher and Kotuby-Amacher 2001). Dunaway et al. (1994) found that data on root-volume density and soil texture were needed to predict streambank soil erodibility. Linking increased soil ero-

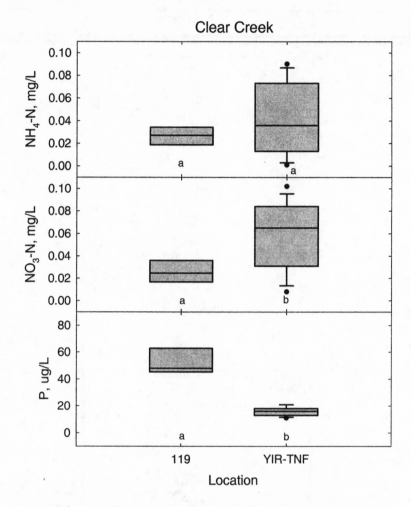

FIGURE 6.8. Box plots of nutrient concentrations at different locations in the Clear Creek and Reese River catchments. The 25th and 75th percentiles are shown as boxes around the 50th percentile, the 10th and 90th percentiles are

sion and sediment transport to land-use activities is made more problematic by the fact that these streams are prone to incision and that their geomorphology is dictated largely by past climate events (chapters 3 and 4).

Despite the ubiquitous recreational and grazing uses of national forest lands in the Monitor, Toquima, and Toiyabe Ranges, nutrient (NH_4-N, NO_3-N, and dissolved o-PO_4) concentrations in streams draining these mountain ranges tended to be low (fig. 6.2). Our results are similar to Tiedemann et al. (1989), who found that stream chemical composition was not affected by increasing intensity of grazing management on eastern Ore-

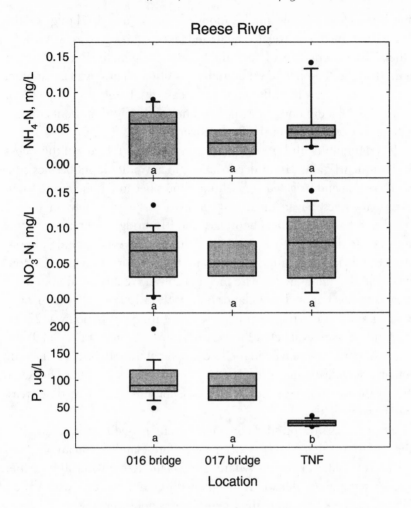

Reese River

shown as error bars, and the 5th and 95th percentiles are shown as points. Water quality parameter means for each catchment location not indicated by the same lowercase letter above each x-axis are significantly different (p < 0.01).

gon rangelands. Tiedemann et al. (1989) concluded that stocking at a rate of 2.8 hectares per animal unit month, even in settings where cattle have access to riparian areas, would not likely have an adverse effect on chemical water quality.

However, agricultural inputs to some of the streams in central Nevada could be detected once they left national forest lands and flowed through adjacent cultivated fields on private lands abutting the streams. For example, increased dissolved P concentrations were found in Clear Creek and the Reese River below irrigated fields and fenced grazing areas (fig. 6.8). Location 119 on Clear Creek and locations 017 bridge and 016 bridge on the Reese River are below irrigated fields and fenced grazing areas, whereas location YIR-TNF on Clear Creek and location TNF on the Reese River are at the Toiyabe National Forest boundaries. Although dissolved constituent concentrations can be influenced by stream discharge (e.g., fig. 6.7 for Birch Creek), it did not significantly influence dissolved nutrient concentrations in Clear Creek or the Reese River.

It is difficult to attribute the increased dissolved P levels in the Reese River at the 017 and 016 bridge locations to any other input besides confined grazing and irrigated agricultural inputs because none of the tributaries to the Reese River contain high levels of dissolved P. Between the national forest and wilderness boundary and 016 bridge, the major streams flowing into the Reese River (and then only in wet years) include Illinois, Stewart, Clear, Marysville, Tierney, and the combined flows of Cottonwood and San Juan Creeks. The mean dissolved P concentration in these streams exiting national forest lands from 1994 to 1998 was 13 micrograms per liter and the maximum value observed during this period was 22 micrograms per liter, well below the mean dissolved P concentration of 98 micrograms per liter observed in the Reese River at the 016 bridge. The only potential significant sources of P to the Reese River once it enters the Reese River Valley are homes, runoff from roads, confined animal feeding areas, and irrigated fields.

Clear Creek's dissolved P concentration increased from a mean of 16 micrograms per liter where it exits national forest land to a mean of 53 micrograms per liter just above where it enters the Reese River. If the other streams accumulate P loads as they flow through the Reese River Valley, then an increased P load to the Reese River is not surprising.

Comparison with Other Water Quality Databases

Water quality databases for streams in central Nevada are limited. The U.S. Geological Survey (USGS) collected water quality data from Stewart Creek in the Toiyabe Range from 1986 through 1991 as part of the hydrologic assessment of the Yucca Mountain nuclear waste storage site (McKinley and Oliver 1994). Its water quality sampling station was 1.3 miles above the Columbine Campground, which is as far upstream as we sampled during our 1995–1998 sampling period. Stream water chemical composition values between the USGS and USFS databases are comparable. However, small but statistically significant lower concentrations of Na, K, Mg, Ca, alkalinity, and Si were found at the USGS site in comparison to the USDA Forest Service site at the Columbine Campground (data not shown). No significant differences between the sites were found for NO_3-N, P, S, F, and Cl. The small differences in some of the constituents between the two sites are attributed to location in the catchment rather than changes in chemical composition with time. Because the Forest Service sampling site is 1.3 miles downstream from the USGS site, slight increases in the concentrations of Na, K, Mg, Ca, alkalinity, and Si can be attributed to the longer flow path and hence time of contact of stream waters with soils and rock formations as the stream flows down the canyon. These are the ions most likely to increase in concentration as stream waters flow through the andesitic rocks, ash-flow tuffs, breccias, lavas, and soils derived from them as waters flow downstream.

The USGS also maintained a Hydrologic Benchmark Network station in the South Twin River catchment. The chemical composition data from our 1994 sampling of the South Twin River are within the range of observed values in the 1963 to 1995 USGS database for this stream, with most values falling between the 25 percent and 75 percent quartiles (Mast and Clow 2000). Mast and Clow (2000) did not find any temporal trends in the chemical composition of the South Twin River except for a slight decrease in the concentration of K with time for which no satisfactory explanation could be found. They concluded that the chemical composition of the South Twin River is controlled by catchment lithology rather than land use, which fits our hypothesis. They also found that rock-weathering-derived constituents (e.g., Na, K, Mg, Ca, alkalinity, Si) were negatively correlated with stream discharge, which they interpreted as weathering-enriched base flow diluted by snowmelt and shallow soil water. This is consistent with our findings of the relationships between chemical composition and stream flow (e.g., fig. 6.7).

Clark et al. (2000) calculated flow-weighted nutrient concentrations for eighty-five relatively undeveloped drainage basins throughout the United States. Nutrient concentrations found in the 1994 survey of streams of the Monitor, Toquima, and Toiyabe Ranges and in the more-detailed 1995–1998 survey of streams in the Toiyabe Range were similar to the flow-weighted concentrations that Clark et al. (2000) derived (table 6.3). Median NH_4-N and dissolved P were slightly higher and median NO_3-N concentrations were slightly lower in central Nevada streams than in the USGS database for undeveloped basins. Higher maximum NH_4-N and

TABLE 6.3.

Comparison of nutrient concentrations (milligrams per liter) from the 1994
survey of Monitor, Toquima, and Toiyabe Range streams and the 1995–1998
survey of Toiyabe Range streams with flow-weighted nutrient concentrations
calculated by Clark et al. (2000) for relatively undeveloped stream basins
throughout the United States

	NH_4-N		NO_3-N		o-PO_4 as P	
Statistic	USGS[1]	USFS[2]	USGS[1]	USFS[2]	USGS[1]	USFS[2]
Minimum	< 0.01	0.000	< 0.01	0.000	< 0.01	0.001
25th percentile	0.016	0.010	0.040	0.045	< 0.01	0.008
Median	0.020	0.029	0.087	0.075	0.010	0.014
75th percentile	0.026	0.046	0.21	0.115	0.011	0.038
Maximum	0.10	0.310	0.77	0.328	0.13	0.400

[1]Flow-weighted nutrient concentrations from Clark et al. (2000)
[2]1994 USDA Forest Service water quality survey of Monitor, Toquima, and Toiyabe Range streams and
1995–1998 USDA Forest Service water-quality survey of Toiyabe Range streams

dissolved P, but lower maximum NO_3-N concentrations, were found in central Nevada streams than in the USGS database for undeveloped basins. The maximum NH_4-N and dissolved P concentrations observed in central Nevada streams were both found in Big Creek, a heavily used recreation canyon. If these outliers are removed from the database, the other statistics remain unchanged, but the maximum NH_4-N and dissolved P concentrations decrease to 0.141 and 0.195 milligrams per liter, respectively. These new maxima for central Nevada streams are much closer to the observed maxima for undeveloped basins in the USGS database.

Effects of Other Factors on Water Quality

Although not considered explicitly in this study, atmospheric deposition can impair catchment water quality via processes such as nitrogen saturation (e.g., Williams et al. 1996) and acid deposition (e.g., Reuss and Johnson 1986). Mast and Clow (2000) compared the chemical composition of South Twin River with volume-weighted mean concentrations of the same constituents in wet-only precipitation collected at the Smith Valley National Atmospheric Deposition Program station located about 180 kilometers southwest of the South Twin River. They found that concentrations of most of the major solutes were much higher in the South Twin River than in precipitation, indicating that rock weathering and not precipitation is the source of most solutes in central Nevada watersheds. The exceptions were NH_4-N, which was higher in precipitation (volume-weighted mean of 0.196 milligrams per liter) than in the South Twin River (maximum of

0.130 milligrams per liter), and NO_3-N, which had similar concentrations in both precipitation (volume-weighted mean of 0.154 milligrams per liter) and the South Twin River (median of 0.090 milligrams per liter, maximum of 0.280 milligrams per liter).

As noted by Clark et al. (2000), atmospheric deposition of N apparently exceeds biological N demand in several forested basins in the eastern United States and Rocky Mountains (Stoddard 1994; Williams et al. 1996) and can account for nearly all the downstream N load in some Midwestern and northeastern streams (Smith et al. 1987; Puckett 1995). The analysis by Mast and Clow (2000) indicates that atmospheric deposition would also account for observed N concentrations in central Nevada streams.

There are several types of riparian plant communities along central Nevada streams, including mixed grass, sedge, and willow-dominated types of communities, depending on the geomorphic setting of the riparian area. Uplands, on the other hand, tend to be dominated by sagebrush and pinyon-juniper plant community associations. Although vegetation community type can potentially influence water quality, we were unable to discern any relationship between water quality parameters and plant community type for central Nevada streams. For example, water quality parameters in the main channel of Washington Creek did not change flowing down valley (data not shown) even though the upper part of the catchment is dominated by pinyon-juniper and sagebrush communities while in the lower part of the catchment the stream flows through several extensive riparian meadows dominated by grasses, sedges, and forbs. There may be some more subtle influences of plant community type on some water quality variables that may be detectable on a seasonal basis, but this would require more extensive site and time-intensive sampling than the resources available for the surveys reported here.

Assessment of Water Quality in the Monitor, Toquima, and Toiyabe Ranges

Streams in the mountain ranges of central Nevada generally have high water quality. None of the streams exceeded current or proposed drinking water standards except for the maximum concentration of Na (28.6 milligrams per liter) in the lower Reese River (016 bridge sampling site), which was more than the 20-milligrams-per-liter health advisory for this element (table 6.4). Trace element concentrations were low despite widespread mineralization and the existence of old mine sites and waste-rock piles throughout these mountain ranges (table 6.4). Central Nevada streams are

TABLE 6.4.

Comparison of maximum concentrations of constituents observed in water samples collected in the 1994 water quality survey of streams in the Monitor, Toquima, and Toiyabe Ranges with selected water quality standards or guidelines

Constituent	National drinking water standards[1]	Proposed national drinking water standards[1]	CASWQCB limits for irrigation water[2]	FAO limits for irrigation water[3]	CASWQCB limits for livestock[4]	Maximum concentration observed	Streams where maximum concentration was observed (Range)
	MCL,[5] mg/L				mg/L		
Primary							
Arsenic	0.05	0.03	1.0	0.10		0.011	Reese River
Cadmium	0.01	0.005		0.01	1	0.002	Jefferson Creek
Chromium	0.05	0.1		0.10	5	0.001	North Twin River
Fluoride	4.0					0.50	Meadow Creek
Lead	0.05					0.005	Danville, Birch, Last Chance, Clear Creeks (Toiyabe)
Nitrate	10.0	10.0		5.0	200	0.629	Kingston Creek
Selenium	0.01	0.05		0.02		0.005	Faulkner, Allison, Clear Creeks (Monitor), Danville, Willow Creeks (south Monitor), Willow Creek (west Monitor)
Secondary							
Aluminum		0.05				<0.05	Willow Creek (west Monitor)
Chloride	250		100	106	1500	9.14	Willow Creek (west Monitor)
Copper	1			0.20		0.002	Meadow Creek
Fluoride	2.0		0.1	1.0	1	0.50	Meadow Creek
Iron	0.3			5.0		0.022	Birch Creek
Manganese	0.05			0.20		0.004	Reese River, Marysville Creek
PH	8.5		8.5	8.5	8.5	8.72	Meadow Creek
Sulfate	250		200		500	65.9	Danville Creek
TDS[6]	500		500	450	2500	313	Danville Creek
Zinc	5.0			2.0		0.009	Big Creek

Constituent	National drinking water standards[1]	Proposed national drinking water standards[1]	CASWQCB limits for irrigation water[2]	FAO limits for irrigation water[3]	CASWQCB limits for livestock[4]	Maximum concentration observed	Streams where maximum concentration was observed (Range)
	MCL[5], mg/L				mg/L		
Health advisory							
Sodium	20					28.59	Reese River
Other							
Bicarbonate				91.5	500	262	Danville Creek
Boron			0.5	0.7		0.1	Willow Creek (south Monitor), Willow Creek (west Monitor), Reese River
Cobalt				0.05		<0.001	Clear Creek (Monitor)
Calcium					500	59.94	Big Creek
Magnesium					250	25.76	Shoshone Creek
Molybdenum				0.01		0.016	Reese River
Nickel				0.20		<0.005	Reese River
SAR[7]			6.0	3		1.11	Reese River
Sodium					1000	28.59	Reese River
RSC[8], meq/L			1.25	1.25		0.73	Reese River

[1]Source: van der Leeden et al. 1990.

[2]Threshold concentrations below which irrigation water should be satisfactory for almost all crops and almost any arable soil. Source: Van der Leeden et al. 1990. CASWQCB = California State Water Quality Control Board.

[3]Concentrations below which there are no restrictions on irrigation water use. Source: Van der Leeden et al. 1990. FAO = Food and Agriculture Organization of the United Nations.

[4]Threshold concentrations below which livestock should show no effects from prolonged use. Source: Van der Leeden et al. 1990.

[5]MCL = maximum contaminant level.

[6]TDS = total dissolved solids.

[7]SAR = sodium absorption ratio = $Na/(Ca + Mg)/2$ where Na, Ca, and Mg concentrations are in meq/L (milliequivalents per liter).

[8]RSC = residual sodium carbonate = $(HCO_3 + CO_3) - (Ca + Mg)$ where the HCO_3, CO_3, Ca, and Mg concentrations are in meq/L.

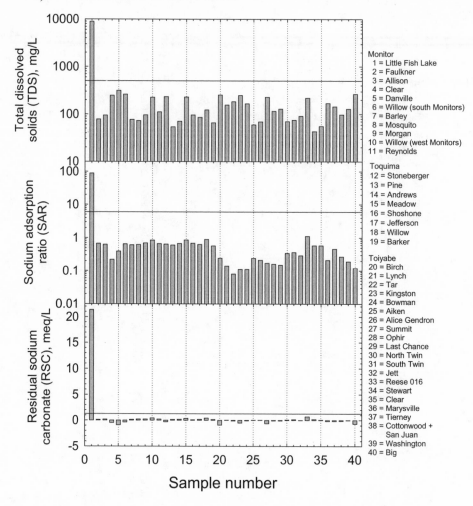

FIGURE 6.9. Bar graphs of selected irrigation water quality parameters for streams in the Monitor, Toquima, and Toiyabe Ranges. The solid horizontal lines indicate recommended limits for each water quality parameter.

also suitable as sources of drinking water sources for livestock. None of the sampled streams exceeded California State Water Quality Control Board (CASWQCB) limits (table 6.4).

The sampled streams generally had a high suitability for irrigation. Total dissolved solids, sodium absorption ratios (SAR), residual sodium carbonate (RSC) levels, and sulfate and chloride levels were all well below recommended irrigation water quality limits (fig. 6.9). Only three of the sampled streams (Danville, Meadow, and Birch Creeks) had pH levels slightly

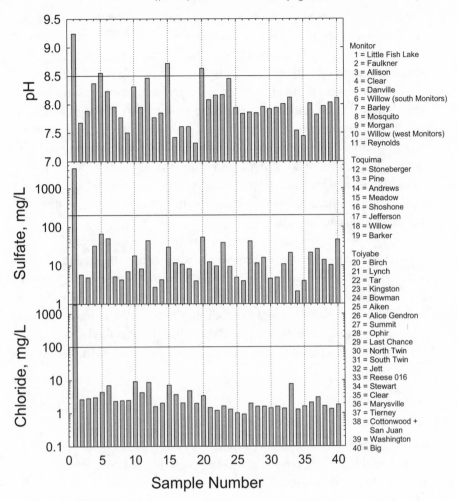

Monitor
1 = Little Fish Lake
2 = Faulkner
3 = Allison
4 = Clear
5 = Danville
6 = Willow (south Monitors)
7 = Barley
8 = Mosquito
9 = Morgan
10 = Willow (west Monitors)
11 = Reynolds

Toquima
12 = Stoneberger
13 = Pine
14 = Andrews
15 = Meadow
16 = Shoshone
17 = Jefferson
18 = Willow
19 = Barker

Toiyabe
20 = Birch
21 = Lynch
22 = Tar
23 = Kingston
24 = Bowman
25 = Aiken
26 = Alice Gendron
27 = Summit
28 = Ophir
29 = Last Chance
30 = North Twin
31 = South Twin
32 = Jett
33 = Reese 016
34 = Stewart
35 = Clear
36 = Marysville
37 = Tierney
38 = Cottonwood +
      San Juan
39 = Washington
40 = Big

higher than the recommended limit of pH 8.5 for irrigation water, though this limitation is not nearly as serious as that of excess levels of salinity and Na. When compared to a saline/sodic lake (Little Fish Lake) found in the desert basin east of the southern part of the Monitor Range, the high irrigation water quality of these streams is even more apparent (fig. 6.9). Because of a lack of outflows and high evaporation, seasonal and permanent desert playa lakes are saline and have high concentrations of Na and other elements that make them wholly unsuited for crop irrigation. Little Fish Lake contains total dissolved solids, SAR, RSC, pH, sulfate, and Cl levels well above the recommended limits for irrigation (fig. 6.9). In addition,

Little Fish Lake contains Mo, B, and F levels above the recommended limits. Because desert basin lakes accumulate salts, the high water quality of streams draining national forest lands makes them a vital resource for economic development and resource sustainability.

The assessment of stream water quality presented here pertains only to the chemical composition of the streams with respect to major and trace inorganic constituents. Since pathogenic viruses, bacteria, and protozoa were not analyzed, no assessment of microbial water quality could be made. Because these are surface waters and are found in areas with livestock grazing and recreational uses, none of these waters can be considered safe sources of potable water without treating to remove or destroy the pathogens prior to use.

Also, these natural waters were not analyzed for industrial chemicals such as pharmaceuticals and hormones as was recently reported in a USGS water quality survey (Kolpin et al. 2002). These chemicals were found to be prevalent in streams selected for their susceptibility to contamination from intense urbanization and livestock production. An analysis of streams in more remote areas with low population densities has not been done, so nothing is known about the prevalence of these chemicals in the broader environment.

Management and Restoration Implications

Because of remoteness and low population density, water quality of streams on national forest lands in central Nevada shows little effects from the relatively light resource-use activities they currently receive. In general, the chemical composition of these stream waters is not sensitive to current land-use activities and chemical water quality has not been impaired. However, because the chemical composition of the stream water is being regulated by catchment lithology and discharge (climate driven), this indicator of water quality may be sensitive to anthropogenic inputs in some catchments should land-use activities change (e.g., changing fire frequency and severity, changing grazing and recreation intensity and duration). Resampling the streams every five to ten years at the base of the catchments is recommended to detect broadscale changes to water quality parameters that apply to the catchment as a whole.

It must be realized that the seasonal sampling reported in this study is too infrequent to detect more-transient water quality impairments. If such impairments are suspected in limited areas because of a shift in land-use activities, then more frequent sampling may be necessary to detect effects from disturbances. For example, Tate et al. (1999) found that where large storm event and seasonal variability were found on California rangeland watersheds, a minimum sampling strategy should include sample collection before, during, and after storms. For Great Basin streams, where snowmelt is the major episodic hydrologic event rather than individual storms, seasonal sampling before, during, and after snowmelt and at base flow should capture all but the most transient variability.

Acknowledgments

We thank Angie Willis, Ezio Buselli, Deb Kutterer, and Tracy Christopherson for assistance with the laboratory analysis. We thank Lisa Stillings, Paul Lechler, and Jerry Miller for their valuable comments on an earlier version of this chapter.

Literature Cited

Amacher, M. C., and J. Kotuby-Amacher. 2001. Stream cutbank erosion in central Nevada streams. In *Sediment monitoring, modeling, and managing.* Proceedings of the 7th Federal Interagency Sedimentation Conference, pp. P1–P4; Mar. 25–29, 2001; Reno, Nev.

APHA, AWWA, WEF (American Public Health Association, American Water Works Association, and Water Environment Federation). 1992. *Standard methods.* 18th ed. Washington, D.C.: APHA, AWWA, WEF.

Bradford, G. R., and D. Bakhtar. 1991. Determination of trace metals in saline irrigation drainage waters with inductively coupled plasma optical emission spectrometry after preconcentration by chelation-solvent extraction. *Environmental Science and Technology* 25:1704–1708.

Clark, G. M., D. K. Mueller, and M. A. Mast. 2000. Nutrient concentrations and yields in undeveloped stream basins of the United States. *Journal of the American Water Resources Association* 36:849–860.

Doner, H. E., and P. R. Grossl. 2002. Carbonates and evaporates. Pp. 199–228 in *Soil mineralogy with environmental applications,* edited by J. Dixon and D. Schultze. Madison, Wisc.: Soil Science Society of America.

Dunaway, D., S. R. Swanson, J. Wendel, and W. Clary. 1994. The effect of herbaceous plant communities and soil textures on particle erosion of alluvial streambanks. *Geomorphology* 9:47–56.

Eaton, A. D., L. S. Clesceri, and A. E. Greenberg, eds. 1995. *Standard methods.* 19th ed. Washington, D.C.: APHA, AWWA, WEF.

Gordon, N. D., T. A. McMahon, and B. L. Finlayson. 1992. *Stream hydrology.* New York: John Wiley & Sons.

Grossl, P. R., and W. P. Inskeep. 1992. Kinetics of octacalcium phosphate crystal growth in the presence of organic acids. *Geochimica et Cosmochimica Acta* 56:1955–1961.

Inskeep, W. P., and P. R. Bloom. 1986. Kinetics of calcite precipitation in the presence of water-soluble organic ligands. *Soil Science Society of America Journal* 50:1167–1172.

Kleinhampl, F. J., and J. I. Ziony. 1985. *Geology of northern Nye County, Nevada.* Bulletin 99A. Nevada Bureau of Mines and Geology, University of Nevada, Reno.

Kolpin, D. W., E. T. Furlong, M. T. Meyer, E. M. Thurman, S. D. Zaugg, L. B. Barber, and H. T. Buxton. 2002. Pharmaceuticals, hormones, and other organic waste-water contaminants in U.S. streams, 1999–2000: A national reconnaissance. *Environmental Science and Technology* 36:1202–1211.

Lachat Instruments. 1991a. *Chloride in waters.* QuikChem Method 10-117-07-1-C. Milwaukee, Wisc.

———. 1991b. *Nitrate/nitrite, nitrite in surface water, wastewater.* QuikChem Method 10-107-04-1-C. Milwaukee, Wisc.

———. 1992. *Alkalinity (methyl orange) in surface and waste waters.* QuikChem Method 10-303-31-1-A. Milwaukee, Wisc.

———. 1993. *Ammonia (phenolate) in potable and surface waters.* QuikChem Method 10-107-06-1-B. Milwaukee, Wisc.

Mast, M. A., and D. W. Clow. 2000. *Environmental characteristics and water quality of Hydrologic Benchmark Network stations in the western United States.* Circular 1173-D. U.S. Geological Survey.

McKinley, P. W., and T. A. Oliver. 1994. *Meteorological, stream-discharge, and water-quality data for 1986 through 1991 from two small basins in central Nevada.* Open-file report 93-651. Denver: U.S. Geological Survey.

Puckett, L. J. 1995. Identifying the major sources of nutrient water pollution. *Environmental Science and Technology* 29:408–414.

Reuss, J. O., and D. W. Johnson. 1986. *Acid deposition and the acidification of soils and waters.* New York: Springer-Verlag.

Roberts, R. J., K. M. Montgomery, and R. E. Lehner. 1967. *Geology and mineral resources of Eureka County, Nevada.* Bulletin 64. Nevada Bureau of Mines and Geology, University of Nevada, Reno.

Smith, R. A., R. B. Alexander, and M. G. Wolman. 1987. Water quality trends in the nation's rivers. *Science* 235:1607–1615.

Stewart, J. H., and E. H. McKee. 1977. *Geology and mineral deposits of Lander County, Nevada.* Bulletin 88. Nevada Bureau of Mines and Geology. University of Nevada, Reno.

Stoddard, J. L. 1994. Environmental chemistry of lakes and reservoirs. Pp. 223–284 in *Advances in chemistry series no. 237,* edited by L. A. Baker. Washington, D.C.: American Chemistry Society.

Stumm, W., and J. J. Morgan. 1996. *Aquatic chemistry,* 3rd ed. New York: John Wiley & Sons.

Tate, K. W., R. A. Dahlgren, M. J. Singer, B. Allan-Diaz, and E. R. Atwill. 1999. On California rangeland watersheds: Timing, frequency of sampling affect accuracy of water-quality monitoring. *California Agriculture* 53:44–48.

Tiedemann, A. R., D. A. Higgins, T. M. Quigley, and H. R. Sanderson. 1989. *Stream chemistry responses to four range management strategies in eastern Oregon.* Research Paper PNW-RP-413. Portland, Ore.: USDA Forest Service Pacific Northwest Research Station.

U.S. EPA (United States Environmental Protection Agency). 1976. *Methods for chemical analysis of water and wastes.* EPA-625-/6-74-003a. Cincinnati: U.S. EPA.

van der Leeden, F., F. L. Troise, and D. K. Todd. 1990. *The water encyclopedia.* 2nd ed. Boca Raton Fla.: Lewis Publishers.

Williams, M. W., J. S. Baron, N. Caine, R. Sommerfeld, and R. Sanford Jr. 1996. Nitrogen saturation in the Rocky Mountains. *Environmental Science and Technology* 30:640–646.

Williams, O. R., R. B. Thomas, and R. L. Daddow. 1988. *Methods for collection and analysis of fluvial sediment data.* WSDG-TP-00012. Washington, D.C.: USDA Forest Service.

Chapter 7

Effects of Geomorphic Processes and Hydrologic Regimes on Riparian Vegetation

Jeanne C. Chambers, Robin J. Tausch,
John L. Korfmacher, Dru Germanoski,
Jerry R. Miller, and David Jewett

The degradation of riparian areas results in changes in the composition, pattern, and areal extent of riparian vegetation (Stromberg 2001). In the central Great Basin, as in other semiarid regions, riparian ecosystems exhibit widespread degradation. Restoration and management of these riparian areas require an understanding of (1) the causes of disturbance and (2) the relationships among riparian vegetation and geomorphic and hydrologic processes (Goodwin et al. 1997). Although the degradation of riparian areas often has been attributed largely to anthropogenic disturbance, in arid and semiarid regions both past and present climate strongly influence geomorphic and fluvial processes and, thus, riparian vegetation. In upland watersheds of the central Great Basin, climate-driven changes in hillslope and fluvial processes that occurred during the mid- to late Holocene are still affecting the composition and·pattern of riparian vegetation over a broad range of scales. Of major significance was a period of extended drought that occurred between 2500 and 2000 YBP. Consequences of the drought include the erosion and depletion of hillslope sediment reservoirs, and the subsequent aggradation of valley bottoms and expansion of side-valley alluvial fans (Miller et al. 2001; chapter 3). The depletion of hillslope sediments has resulted in streams that are currently sediment limited and, thus, have a natural tendency to incise. The expansion of side-valley alluvial fans has resulted in stepped valley profiles that often are related to abrupt changes in the morphology, hydrology, and vegetation of associated valley segments and stream reaches. The rate, magnitude, and pattern of stream incision differ among watersheds and depend

196

largely on basin sensitivity to disturbance as governed by factors such as geology, basin relief and morphometry, and valley width and gradient as well as by the influence of side-valley alluvial fans (chapter 4).

In many cases, anthropogenic disturbances such as roads in the valley bottoms, overgrazing by livestock, and recreational activities have accelerated the degradation of stream and riparian ecosystems and altered riparian vegetation (Belsky et al. 1999; Martin and Chambers 2001a; Martin and Chambers 2002). Incision and other types of channel change occur during high stream flows and can result from natural disturbances that increase runoff, such as wildfire in the uplands (Germanoski and Miller 1995) and rain on snow (Germanoski et al. 2001). However, incision also can result from anthropogenic disturbances that concentrate stream flows and increase erosional power, such as roads in the valley bottoms (Lahde 2003). Consequences of stream incision include altered channel pattern and form (Miller et al. 2001; chapter 3), changes in surface water and groundwater interactions (chapter 5), lowered water table depths and, ultimately, changes in vegetation composition, pattern, and extent (Wright and Chambers 2002; chapter 5). Anthropogenic disturbances cause changes in stream channels and riparian ecosystems that are in addition to stream incision, including compacted soils and decreased infiltration, altered biogeochemical cycles, and changes in plant physiological, population, and community processes (Belsky et al. 1999; Martin and Chambers 2001a, 2002).

Riparian meadow complexes often occur in incision-dominated watersheds and are susceptible to stream incision (see chapter 5). They also are highly valued for ecosystem services such as forage for native herbivores and livestock and recreational activities and, thus, are frequently overutilized. Because of ongoing degradation in these ecosystems, they are one of the highest priorities for management and restoration.

In this chapter, the relationships among riparian vegetation and geomorphic and hydrologic processes in central Great Basin watersheds are evaluated over a range of scales. These relationships are examined through a series of case studies that have been conducted by the Great Basin Ecosystem Management Project. First, the effects of differences in the geologic and hydrologic characteristics of the watersheds (i.e., basin sensitivity to disturbance) on the composition and pattern of streamside vegetation are investigated. Second, the influence of side-valley alluvial fans on riparian vegetation composition and pattern within riparian corridors is evaluated. Third, relationships among water table regimes, riparian soils, and riparian vegetation composition and dynamics are examined at the scale of the valley segment or stream reach with an emphasis on meadow complexes. The effects of anthropogenic disturbance on meadow complexes also are evaluated at the valley segment or stream-reach scale with a focus on livestock grazing. The chapter concludes with a discussion of the implications for management and restoration.

Composition and Pattern of Streamside Vegetation

Differences in geologic and hydrologic characteristics often exist among watersheds in semiarid regions, and these differences influence the composition and pattern of riparian vegetation at multiple scales (Harris 1988; Bendix 1999; Wasklewicz 2001). In the central Great Basin, watersheds with varying geologic and hydrologic characteristics, including basin relief and morphometry, have been characterized according to basin sensitivity to natural and anthropogenic disturbance (chapter 4).

Influence of Geologic and Hydrologic Characteristics of Watersheds

In this section, the combined influences of watershed characteristics and ongoing stream incision on riparian vegetation are evaluated. Specifically, relationships among the hydrogeomorphic characteristics of the stream channels and the composition and pattern of riparian vegetation are examined for watersheds with different sensitivities to disturbance.

WATERSHED CHARACTERISTICS

This research focused on gaged watersheds representing both flood-dominated (Group 1) and incision-dominated basins (Groups 2 and 4) (see chapter 4) (table 7.1). Group 1 basins are underlain by Tertiary volcanic rock and are characterized by high-relief basins with moderate to high stream power. These are narrow, bedrock-controlled systems with minimal sediment storage. During large floods, significant cutting results in multiple, discontinuous terraces. Group 2 basins are underlain primarily by Tertiary volcanic rocks and have large, high-relief basins and relatively high stream power. The streams have eroded completely through existing side-valley alluvial fans; they have high incision values and relatively smooth longitudinal profiles. These basins appear to be reaching an equilibrium state because their channel morphologies are currently somewhat stable. Group 3 basins are dominated by side-valley alluvial fans and are described in the following section. Group 4 basins are underlain by intrusive igneous and sedimentary rocks and tend to have lower relief and stream power than other basin types. Group 4 basins are pseudostable. They have stream channels that typically exhibit minimal downcutting but have the potential for rapid and catastrophic incision on small or local scales.

METHODS

Valley segments were selected that had relatively uniform characteristics in terms of geology, valley morphometry, stream channels, and vegetation, were located at the base of upland watersheds, and were moderately to highly incised. Each of the valley segments was coincident with a stream gaging station and ranged from 1.5 to 3.0 kilometers in length. Sampling was conducted along five to seven cross-sectional valley transects. The analysis focused on channel parameters because in other semiarid systems, changing channel morphologies and their associated fluvial landforms have greater influence on riparian vegetation distributions than flood hydraulics (Wasklewicz 2001). Sampling methods followed protocols detailed in Davis (2000). Transects were surveyed with a total station. The geomorphic variables measured included valley width, channel slope, number of inset terraces, terrace height above the channel bed, bankfull channel depth, bankfull width/depth ratio, depth of the incised channel, width of the incised channel, channel bed particle size (D_{50}), and percentage of bank particles less than 2 millimeters in size. Channel bed particle size (D_{50}) was determined along a 30-meter stretch of stream using methods modified from Wolman (1954). Bank particle size was obtained by sieving four bulked samples that were collected from the right and left channel banks 5 meters upstream and downstream of the transects.

Vegetation was sampled along the same transects as the geomorphic variables. Nested frequency of herbaceous vegetation was determined for each stream terrace on both the left and right banks from three 0.1-square-meter quadrats placed 2.0 meters apart (Castelli et al. 2000). Stem density of woody vegetation was recorded at the same locations from three 1.0-square-meter quadrats. Relative stem density was recorded for clonal species (*Salix* spp., *Rosa woodsii*) using the following scheme: 1 = 1–10; 2 = 11–25; 3 = 26–50; 4 = 51–100; 5 = 101–150; 6 = 151 or greater). Actual stem density was recorded for nonclonal species (*Artemisia*, *Chrysothamnus*, tree species). Tree densities were determined for both banks within 2-meter-wide belts at the center of each transect.

Two-way indicator species analysis (TWINSPAN, Hill 1979; McCune and Medford 1999) was used to classify vegetation samples into vegetation

TABLE 7.1.

Locations and characteristics of the study watersheds

Drainage	Latitude/Longitude of Gaging Station	Reference Elevation (m)	Basin Geology	Basin Area (km²)	Mean Annual Discharge (m³s)	Mean Peak Discharge (m³s)
Group 1 **Flood Dominated**						
South Twin River	38°53′15″N 117°14′40″E	1,950	Volcanic: welded tuff, andesite	52.5	0.199 ± 0.022	1.762 ± 0.416
Pine Creek	38°47′40″N 116°51′13″E	2,300	Volcanic: rhyolite, andesite	32.0	0.164 ± 0.021	1.919 ± 0.465
Mosquito Creek	38°48′22″N 116°40′43″E	2,195	Volcanic: welded tuff	39.6	0.070 ± 0.013	0.644 ± 0.178
Group 2 **Deeply Incised**						
Upper Reese River	38°51′00″N 117°28′00″E	2,165	Volcanic: andesite, basalt	139.1	0.353 ± 0.051	4.321 ± 1.026
Lower Kingston Creek	39°12′45″N 117°06′45″E	1,975	Sedimentary: shale, limestone	61.4	0.270 ± 0.029	1.523 ± 0.414
Currant Creek	38°50′50″N 115°22′00″E	2,040	Volcanic/ sedimentary	33.9	0.096 ± 0.018	2.076 ± 1.057
Group 4 **Pseudostable**						
Illipah Creek	39°19′07″N 115°23′39″E	2,085	Sedimentary	82.7	0.093 ± 0.033	1.755 ± 1.140

Stream flow data from U.S. Geological Survey gaging stations (http://water.usgs.gov/nv/nwis/sw)

types. The analyses used proxy frequency data based on presence/absence of herbaceous and shrubby species within the three subsamples at each well location (D. Weixelman, personal communication). Species that occurred in one subsample were assigned a value of 1, those that occurred in two subsamples were given a value of 2, and species that were present in all three subsamples were given a value of 3. Canonical correspondence analysis (CCA) was used to evaluate relationships among the vegetation types and the geomorphic variables (ter Braak 1986). *Artemisia tridentata* vegetation types were eliminated from the CCA as these types consistently occurred on the highest-elevation terraces (valley floor), and their occurrence within a valley segment was largely unrelated to channel attributes.

RELATIONSHIPS AMONG GEOMORPHIC CHARACTERISTICS AND VEGETATION TYPES

Various studies have found riparian plant species and communities to be related to landform type and position, substrate characteristics, water availability, and tolerance to flooding (see reviews in Hupp and Osterkamp 1996; Hughes 1997; Bendix and Hupp 2000). In this study, the vegetation types identified included *Carex nebrascensis* meadow, mesic meadow, *Salix* spp./mesic meadow, *Salix* spp./mesic forb, *Betula occidentalis*/mesic meadow, *Artemisia tridentata tridentata*/*Leymus cinereus*, and *Artemisia tridentata tridentata*/*Poa secunda*. These types correspond roughly to the ecological types described by Weixelman et al. (1996) in a regional analysis of central Nevada riparian areas (appendix 7.1). Two additional types also were identified: dense *R. woodsii* and *Prunus virginiana*/*R. woodsii*. Eight geomorphic variables were significantly related to the vegetation types on one or both axes in the CCA (fig. 7.1). The eigenvalues for axis 1 and axis 2 were 0.350 and 0.289, respectively, and the correlation of vegetation type with the combined environmental variables was about 80 percent for both axes. The correlation coefficients from the CCA indicated that channel particle D_{50}, terrace height, width/depth ratio, channel slope, bank particle size (percentage less than 2 millimeters), incised channel depth, number of terraces, and bankfull depth were related to vegetation type on one or both axes ($R^2 = 0.58$ to 0.15; $P < 0.05$ to $P < 0.01$).

These results are consistent with those from other arid and semiarid areas and indicate that the vegetation types are closely related to the hydrologic and geomorphic characteristics of the study watersheds. At local scales, water availability, as indicated by surface elevation above the water surface or stream channel, is often the primary control on riparian species

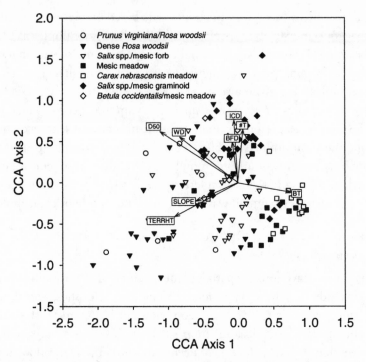

FIGURE 7.1. Relationships among stream geomorphic characteristics and vegetation types for the study watersheds as indicated by Canoncal Correspondence Analyses (CCA). Arrows indicate the direction and approximate magnitude of effects for environmental variables in the model.

SLOPE = channel slope; TERRHT = terrace height; BFD = bankfull channel depth; WD = bankfull width/depth ratio; ICD = incised channel depth; D_{50} = channel particle size (D_{50}); BT = percentage bank particle size less than 2 millimeters; and #T = number of terraces.

distributions (Furness and Breen 1980; Hughes 1990; Hupp and Osterkamp 1996; Stromberg et al. 1996; Merritt and Cooper 2000; Wasklewicz 2001). Terrace height was more closely related to vegetation type than any other variable in the study watersheds. Because terrace height is correlated with a suite of environmental variables that are influenced by the hydrologic regime, including shear stress, sediment deposition and erosion, soil water, and soil oxygen concentration, it is often correlated with the distributions of riparian species (Auble et al. 1994; Merritt and Cooper 2000). Channel and bank particle size also were highly correlated with vegetation type in the study watersheds. Similarly, for the San Pedro River in Arizona, depth to groundwater and its spatial correlate, geomorphic surface or terrace elevation, had the greatest influence on species composition, followed by soil texture and moisture-holding capacity (Stromberg et al. 1996).

At larger scales, species occurrence and community composition is affected not only by the hydrologic regime and water availability, but also by fluvial geomorphic processes and landforms (Harris 1988; Bendix 1999; Wasklewicz 2001). Thus, a larger set of stream channel attributes is necessary to define the physical setting of vegetation types that occur in watersheds or valley segments with varying geomorphic characteristics. For the study watersheds, channel variables that were significantly associated with vegetation types, including those that influence water availability (table 7.2), can be used to characterize the geomorphic setting of the vegetation types. Although some of the vegetation types are broadly distributed among the watersheds, others occur in only one or two watersheds, reflecting more narrowly defined geomorphic settings.

The *C. nebrascensis* meadow type occurs only on the lowest stream terraces (0.38 meters ± 0.06) and is composed of obligate and facultative wetland species (USDA NRCS 2002) such as *C. nebrascensis, Deschampsia cespitosa,* and *Juncus balticus* (table 7.2; appendix 7.1). It is typically associated with low-gradient streams with small channel particle sizes, fine-textured banks, and low width/depth ratios (table 7.2). The other grass- and sedge-dominated type, mesic meadow, occurs on slightly higher terraces (0.87 ± 0.16 meter) and is characterized by facultative wetland species such as *Poa pratensis, Elymus trachycaulus, J. balticus,* and *Aster occidentalis* (table 7.2; appendix 7.1). Channel characteristics for the mesic meadow vegetation type are moderate stream gradients and, like the *C. nebrascensis* vegetation type, relatively small channel and bank particle sizes and low width/depth ratios (table 7.2). Riparian wet meadow communities of sedges and rushes (*C. nebrascensis, J. balticus*) have two to six times the root density and biomass of grasses like *P. pratensis* and *D. cespitosa* (Manning et al. 1989; Dunaway et al. 1994). Consequently, sedges and rushes provide greater resistance to erosion (Dunaway et al. 1994) and compression (Kleinfelder et al. 1992) and are more effective at stabilizing stream channels than are riparian grasses. However, because riparian obligate sedges and rushes require relatively high water tables, they can rapidly decrease in abundance following stream incision.

The *Salix* spp./mesic meadow type occurs on terraces that average 0.49 ± 0.05 meter in height, while the *Salix* spp./mesic forb type is found on terraces 0.73 ± 0.11 meters high (table 7.2; appendix 7.1). In the study systems, the latter type consistently has a significant *R. woodsii ultramontana* component and frequently has a minor component of *Populus tremuloides, P. angustifolia,* or *P. balsamifera trichocarpa.* Relative to the other types,

TABLE 7.2.

Geomorphic characteristics of the dominant vegetation types within the study watersheds

Vegetation Type	Terrace Height (m)	Channel Particle D_{50}	Bank Particles <2 mm (%)	Slope (%)	Channel Width/depth Ratio	Bankfull Depth (m)	Number of Terraces	Incised Channel Depth (m)
Artemisia tridentata tridentata/ Leymus cinereus or *Poa secunda*	1.85 ± 0.19	39.2 ± 4.4	60.5 ± 5.6	0.035 ± 0.006	6.12 ± 0.57	0.59 ± 0.09	1.80 ± 0.13	1.92 ± 0.22
Prunus virginiana/Rosa woodsii	1.77 ± 0.40	59.7 ± 5.3	26.4 ± 6.4	0.054 ± 0.014	9.26 ± 1.48	0.50 ± 0.09	1.50 ± 0.22	1.38 ± 0.37
Dense *Rosa woodsii*	1.56 ± 0.18	48.9 ± 3.4	50.7 ± 5.7	0.041 ± 0.007	7.87 ± 0.63	0.49 ± 0.27	1.77 ± 0.17	1.95 ± 0.25
Betula occidentalis/mesic meadow	1.04 ± 0.40	65.4 ± 3.0	35.0 ± 11.4	0.048 ± 0.008	7.38 ± 1.54	0.84 ± 0.29	1.75 ± 0.16	1.63 ± 0.31
Mesic meadow	0.87 ± 0.16	25.5 ± 6.4	60.9 ± 7.5	0.034 ± 0.008	4.90 ± 0.63	0.52 ± 0.03	1.92 ± 0.15	1.57 ± 0.14
Salix spp./mesic forb	0.73 ± 0.11	55.3 ± 3.1	52.8 ± 6.3	0.047 ± 0.006	6.60 ± 0.48	0.53 ± 0.04	1.63 ± 0.15	1.45 ± 0.15
Salix spp./mesic meadow	0.49 ± 0.05	51.7 ± 5.5	60.8 ± 7.7	0.034 ± 0.009	8.02 ± 0.82	0.53 ± 0.04	2.00 ± 0.30	2.48 ± 0.47
Carex nebrascensis meadow	0.38 ± 0.06	15.3 ± 5.0	77.6 ± 6.0	0.018 ± 0.006	4.48 ± 0.41	0.50 ± 0.02	1.78 ± 0.15	1.50 ± 0.19

the *Salix* vegetation types are associated with moderate to high stream gradients, large channel particle sizes and fairly fine bank textures (table 7.2). *Salix* spp. are densely rooted and are highly effective at stabilizing the lower terraces of stream channels (Thorne 1990). Both *Salix* spp. and *Populus* spp. depend largely on flood events for regeneration. However, *Salix* spp. are typically less drought tolerant but more flood tolerant than *Populus* spp. (Van Splunder et al. 1996; Amlin and Rood 2002) and generally occur at lower terrace elevations and closer to the stream than *Populus* spp. (Busch et al. 1992; Amlin and Rood 2002). In the study watersheds, *Salix* occurs primarily on lower terraces (table 7.2), while *P. tremuloides*, *P. angustifolia*, and *P. balsamifera* occur on all terraces but are most abundant on intermediate terraces. As might be expected for either incising or flood-dominated systems, increasingly older individuals of the *Populus* species occur on progressively higher terraces within the study systems (J. Chambers and D. Henderson, unpublished data) and elsewhere (Scott et al. 1996).

   The *B. occidentalis*/mesic meadow, dense *R. woodsii*, and *Prunus virginiana*/*R. woodsii* types are located on progressively higher terraces (1.04 ± 0.40, 1.56 ± 0.18, and 1.77 ± 0.40 meters, respectively) (table 7.2; appendix 7.1). These types are associated with steep stream gradients and large channel particle sizes (table 7.2). The *Betula* and *Prunus* types also have coarse bank textures. The dense *Rosa* type is widespread among the watersheds and tends to have intermediate channel characteristics. *Rosa woodsii* is categorized as a disturbance-adapted species (Manning and Padgett 1992; Weixelman et al. 1996) that is abundant in flood-dominated or grazed systems.

   Artemisia tridentata vegetation types (*A. tridentata tridentata*/*L. cinereus*, and *A. tridentata tridentata*/*Poa secunda*) are often the dominant vegetation types on the valley floors and, thus, occur on the highest stream terraces within the riparian corridor (table 7.2; appendix 7.1). Exceptions are in highly confined or flood-dominated systems where riparian shrub and tree species tend to dominate. The *Artemisia* types occur on deep soils with higher available moisture than adjacent hillslopes (Weixelman et al. 1996). Herbaceous species composition of these types is controlled largely by depth to the water table (Wright and Chambers 2002).

Associations among Vegetation Types and Basin Groups

Like other semiarid ecosystems, watersheds with different geomorphic characteristics and sensitivities to disturbance are characterized by unique vegetation associations (Harris 1988). Flood-dominated systems with volcanic lithologies (Group 1), including South Twin River, Pine Creek, and Mosquito Creek, are dominated largely by flood-tolerant species. The *Salix* spp./mesic forb type (with *R. woodsii ultramontana* or *Populus* spp.) and dense *Rosa* types occur on low to intermediate terraces, while *Artemisia* vegetation types are found on upper terraces. Geomorphic characteristics of these systems, and predictive variables for the streamside vegetation types, are high stream gradients, large channel particle sizes, intermediate stream width/depth ratios and incised channel depths (table 7.3).

Incised basins with volcanic lithologies (Group 2), including Upper Reese River, Lower Kingston Canyon, and Currant Creek, have significantly incised channels with trenched side-valley alluvial fans (see chapter 4). Common elements in these systems include lower terraces dominated by *Salix* spp./mesic meadow or *Salix* spp./mesic forb vegetation types, intermediate terraces characterized by *R. woodsii ultramontana* associated vegetation types and upper terraces with *Artemisia* vegetation types. Lower Kingston Creek also supports two other woody vegetation types on intermediate terraces, *B. occidentalis*/mesic meadow and *Prunus virginiana/R. woodsii*. These watersheds have relatively low to intermediate gradients, high numbers of terraces, and high, incised channel depths (table 7.3). Upper Reese River is characterized by the lowest gradient reaches and the finest-textured banks. These characteristics favor graminoid understories in central Great Basin *Salix* spp. vegetation types (Weixelman et al. 1996). In contrast, higher stream gradients, poorly defined lower terraces, and coarser-textured bank soils, like those found in Lower Kingston Canyon and Currant Creeks, favor forb understories.

Incision-dominated basins with crystalline and sedimentary lithologies (Group 4), like Illipah Creek, typically exhibit minor to moderate stream incision. Valley segments with deep alluvium and elevated water tables have the potential for catastrophic incision due to processes associated with groundwater sapping (Germanoski et al. 2001; chapters 4 and 5). Where incised, these streams often have lower and intermediate terraces dominated by the *C. nebrascensis* and mesic meadow types, respectively. Upper terraces are again characterized by *Artemisia* vegetation types. In general, the stream channels exhibit well-defined terraces, low stream gradients, small channel substrates, fine-textured banks and low width/depth ratios (table 7.3).

Influence of Side-Valley Alluvial Fans

Alluvial fans that prograde from tributary valleys into the axial drainage can have major influences on stream processes, hydrologic regimes and, thus,

TABLE 7.3.

Geomorphic characteristics and dominant vegetation types of the study watersheds

Drainage	Dominant Vegetation Types (%)	Channel Slope (m/m)	Number of Terraces	Channel Particle D$_{50}$ (mm)	Bank Particles > 2 mm (%)	Bankfull Width/Depth Ratio	Bankfull Channel Depth (m)	Incised Channel Depth (m)	Incised Channel Width (m)	Valley Width (m)
Group 1 Flood Dominated										
South Twin River	Mesic meadow(31)									
	Artemisia (23)									
	Salix spp./mesic forb (15)									
Pine Creek	Dense Rosa (15)	0.054 ± 0.01	1.8 ± 0.3	53.7 ± 3.9	65.6 ± 3.9	6.93 ± 0.87	0.57 ± 0.06	1.65 ± 0.13	11.42 ± 1.9	23.4 ± 3.6
	Dense Rosa (39)	0.076 ± 0.016	1.2 ± 0.4	59.8 ± 3.0	59.7 ± 9.8	9.00 ± 1.66	0.47 ± 0.08	1.36 ± 0.18	8.58 ± 2.02	46.2 ± 4.9
	Salix spp./mesic forb (17)									
Mosquito Creek	Salix spp./mesic forb (54)									
	Dense Rosa (21)	0.046 ± 0.012	1.4 ± 0.2	49.5 ± 7.2	41.3 ± 10.1	5.70 ± 0.41	0.44 ± 0.04	0.80 ± 0.04	4.82 ± 0.60	46.1 ± 4.6
	Artemisia (21)									
Group 2 Deeply Incised										
Upper Reese River	Salix spp./mesic med (50)									
	Dense Rosa (21)									
	Artemisia (18)	0.014 ± 0.002	2.3 ± 0.3	50.8 ± 5.2	25.1 ± 4.8	8.87 ± 0.92	0.53 ± 0.03	3.13 ± 0.66	19.90 ± 4.69	108.1 ± 28.5
Lower Kingston Creek	Betula/Cornus (32)									
	Salix spp./mesic forb (16)	0.048 ± 0.009	1.7 ± 0.2	67.5 ± 3.9	78.1 ± 3.8	5.93 ± 1.38	1.01 ± 0.31	1.81 ± 0.37	11.03 ± 2.41	54.2 ± 2.9
	Prunus/Rosa (16)									
	Artemisia (16)									
Currant Creek	Artemisia (50)									
	Dense Rosa (31)	0.03 ± 0.005	1.8 ± 0.4	36.9 ± 3.5	58.4 ± 5.6	6.82 ± 0.99	0.44 ± 0.05	2.02 ± 0.13	12.6 ± 2.42	No data
	Salix spp./mesic forb (14)									
Group 4 Pseudostable										
Illipah Creek	Artemisia (50)									
	Mesic meadow (31)									
	C. nebrascensis med (29)	0.012 ± 0.002	1.9 ± 0.1	8.3 ± 2.3	15.4 ± 3.7	4.01 ± 0.38	0.47 ± 0.02	1.49 ± 0.22	8.44 ± 1.25	No data

For complete vegetation type names and dominant species see appendix 7.1

riparian vegetation patterns (e.g., Grant and Swanson 1995; Swanson et al. 1998). Side-valley alluvial fans are dominant features of many upland watersheds in the central Great Basin. The fans reached their maximum extent during the drought that occurred approximately 2500 and 2000 YBP when many of the fans extended across the valley floors (Miller et al. 2001; chapter 3). The fans have resulted in stepped valley profiles and are often associated with major changes in the morphology and hydrology of associated valley segments. Because side-valley alluvial fans influence both geomorphic characteristics and water availability, they affect the types and patterns of riparian vegetation above (upstream), at, and below (downstream) fan deposits.

Watersheds with the most highly developed side-valley alluvial fans often are elongated in planform and are underlain by sedimentary and metasedimentary rocks. These basins exhibit incision-dominated responses (Group 3; see chapter 4), but reaches upstream of the fans often vary considerably with respect to erosional and depositional processes and in the degree of fan entrenchment. Watersheds with volcanic lithologies that exhibit incision-dominated responses (Group 2) also are elongated in planform and characterized by side-valley alluvial fans, but in contrast to Group 3 streams, more of the side-valley alluvial fans have been trenched. The influence of alluvial fans on the geomorphic characteristics and vegetation patterns of associated valley segments and stream reaches has been examined for watersheds in the central Great Basin representing Group 2 basins (Washington Creek, Cottonwood Creek, and San Juan Creek) and Group 3 basins (Big Creek and upper Kingston Creek) (Korfmacher 2001).

METHODS

Study methods are detailed in Korfmacher (2001). Fifty-five cross sections were located for twenty-one alluvial fans that represented three geomorphic positions (above, at, and below fans) and all likely vegetation types. Data on geomorphic parameters likely to influence vegetation patterns were collected for each transect including valley width, slope perpendicular to the channel, channel slope, number of terraces, bankfull width, depth, width/depth ratio, depth of entrenchment, and entrenchment ratio (entrenchment depth/bankfull depth). Vegetation in the study watersheds was classified and mapped from low-altitude, high-resolution multispectral videoimagery using methods described in Neale (1997). Vegetation sampling was conducted from the classified imagery using a GIS (geographic information system) (Korfmacher 2001). A 20 × 24-meter sampling grid was centered on each cross section with the sides parallel to the stream channel. Percent cover class was determined from eight 3-meter-wide zones, four on each side of the stream (0–3, 3–6, 6–9, and 9–12 meters). A split-plot ANOVA model was used to evaluate differences in the geomorphic variables and percentage cover of each vegetation class for fan position and distance from the stream (SAS Institute 2000).

INFLUENCE OF SIDE-VALLEY ALLUVIAL FANS ON GEOMORPHIC CHARACTERISTICS

As expected, alluvial fans influence the geomorphic characteristics of the study watersheds. Three geomorphic variables exhibited significant differences for above, at, and below fan positions—valley width, slope perpendicular to the channel, and bankfull channel width (Korfmacher 2001). Valley width was greatest above fans, intermediate below fans, and least at fans. Also, slope perpendicular to the channel was three times steeper at fans than above or below fans, and bankfull depth was greater at fans.

INFLUENCE OF SIDE-VALLEY ALLUVIAL FANS ON VEGETATION CHARACTERISTICS

The vegetation types show distinct differences with respect to both fan position and distance from the stream (fig. 7.2; Korfmacher 2001). At the fans, *Salix* spp. or *P. tremuloides* are the major streamside components, while below the fans *Salix* spp. and *B. occidentalis* are the dominant streamside components (fig. 7.3). Woody riparian types (*Salix* spp., *P. tremuloides*, *B. occidentalis*) at and below fans are most abundant at 0–3 meters from the stream (greater than 80 percent in many cases), and rapidly decrease in abundance with increasing distance from the stream. Upland vegetation types are also most abundant at and below the fans but increase in abundance with increasing distance from the stream. For upland vegetation types at and above fans, vegetation cover increases from less than 20 percent cover at 0–3 meters to greater than 50 percent at 9–12 meters. Above-fan locations have significantly higher percentages of *C. nebrascensis* and mesic meadow vegetation types than either at-fan or below-fan locations (fig. 7.3). Woody riparian types, *P. tremuloides*, *Salix* spp., and *B. occidentalis*, also are abundant in above-fan locations. In comparison to at-fan and below-fan positions, woody riparian types above fans generally exhibit significantly smaller declines in abundance with distance from the stream. In above-fan positions, the woody riparian types have greater than 35 percent cover at 9–12 meters from the stream.

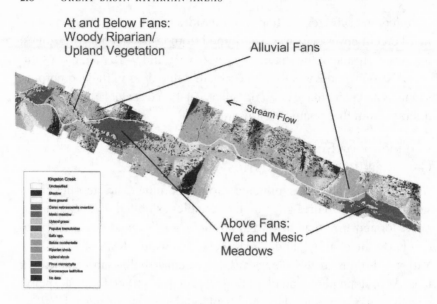

FIGURE 7.2. Vegetation types in relation to alluvial fan position as determined from low-altitude, high-resolution, multispectral videoimagery for a fan-dominated basin (Group 3; Upper Kingston Canyon). Vegetation was classified and mapped using methods described in Neale (1997).

The *C. nebrascensis*, mesic meadow and woody riparian vegetation types all require relatively high water availability. *Carex nebrascensis* and the dominant willow species, *Salix exigua*, and *S. lutea*, are wetland obligate species, *B. occidentalis* is a facultative wetland species, and *P. tremuloides* is a facultative species. The abundance of the *C. nebrascensis*, mesic meadow, and woody riparian vegetation types at all distances from the stream indicates that above-fan locations generally have higher water tables and wider riparian zones than at-fan or below-fan locations.

The prevalence of riparian meadow or woody vegetation types in above-fan locations appears to vary with stream type. The *C. nebrascensis* meadow type is best developed in Group 3 basins with sedimentary and metasedimentary lithologies (Group 3), while woody riparian types, especially *P. tremuloides*, are most abundant in basins with volcanic lithologies (Group 2). The differences in vegetation types may be explained largely by less fan incision and higher water table depths above fans in Group 3 basins, although varying groundwater hydrology due to different lithology also may play a role. Both *B. occidentalis* and *P. tremuloides* occur on higher stream terraces than *C. nebrascensis* (table 7.2), and *P. tremuloides* is capable of

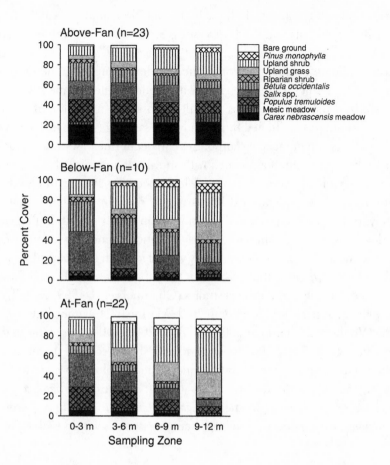

FIGURE 7.3. Mean percentage areal cover by category for different alluvial fan positions and distances from the stream. The cover categories were obtained from classified and mapped low-altitude, high-resolution videoimagery using a GIS (geographic information system).

root sprouting. These traits should favor the persistence of the latter species following stream incision and decreases in water tables.

Dynamics of Riparian Meadow Complexes

In upland watersheds of the Great Basin, meadow complexes often occur in incision-dominated basins—such as basins dominated by side-valley alluvial fans—and are highly susceptible to stream incision. In basins that are dominated by side-valley alluvial fans (Group 3), meadow complexes are located primarily upstream of side-valley alluvial fans (Miller et al.

2001; see chapters 4 and 5). In these basins, incision often starts at the fans as channel gradients and fluid shear stress are at maximum values where axial channels traverse the fans. Once the fans are breached, the wave of incision migrates upstream through the meadow complex in the form of knickpoints or knick zones. In basins with pseudostable channels (Group 4), localized incision occurs as a result of upstream migration of knickpoints and often is facilitated by groundwater sapping (see chapters 4 and 5). Rapid incision of meadow complexes can occur due to short-term forcings such as high runoff associated with snowmelt and rainfall (see chapters 4 and 5). Riparian meadow complexes occur in areas with elevated groundwater and typically exhibit strong hydrologic gradients related largely to topographic position. Stream incision usually results in progressive lowering of water tables and the degradation of meadow complexes. Because meadow complexes are highly valued for both landscape diversity and ecosystem services, they have been the subject of several Great Basin Ecosystem Management Project studies (Chambers et al. 1999; Castelli et al. 2000; Chambers and Linnerooth 2001; Martin and Chambers 2001a,b, 2002; Wright and Chambers 2002; Wehking 2002). Meadows at low to intermediate elevations (2,000 to 2,350 meters) have received the most attention. The common vegetation types along the hydrologic gradient within these meadow complexes are, from wettest to driest, *C. nebrascensis* meadow, mesic meadow, dry meadow, and *Artemisia tridentata tridentata/ L. cinereus* (see appendix 7.1 for representative species within each type).

Hydrologic Regimes

Montane meadow complexes often exhibit high spatial and temporal variability in groundwater regimes (Allen-Diaz 1991; Castelli et al. 2000; chapter 5). In floodplain aquifers, subsurface hydrology is a consequence of flood frequency and duration, flow exchange between the stream channel and the floodplain (hyporheic flow), permeability and heterogeneity of the alluvial substrates, subsurface flows from upland slopes and the contribution from regional aquifers (Bencala 1993; Stanford and Ward 1993; Huggenberger et al. 1998). Large differences in the depth to water table can occur over small distances due to differences in the sources and directions of subsurface flows and in the hydraulic conductivity of the substrate. Also, in montane meadows like those in the central Great Basin, the hydrologic cycle is heavily influenced by snowpack, and summers are typically characterized by drought. Consequently, there can be significant fluctuations in water-table levels both during growing seasons and among years.

Soil Characteristics

The effects of groundwater regimes are often predictable for wetland soils (Mitsch and Gosselink 1993) and are fairly consistent for meadows in the central Great Basin. A comparison of soil physical and chemical characteristics was made for meadow sites that were (1) dominated by *C. nebrascensis* and had consistently high water tables (0–20 centimeters measured over three years in mid-August), (2) codominated by *C. nebrascensis* and mesic graminoid vegetation and had intermediate water tables (−30 to −50 centimeters), and (3) dominated by mesic graminoid vegetation with relatively low water tables (−60 to −80 centimeters) (Chambers et al. 1999). Soil types for the high- to low-water-table sites were, respectively, typic cryaquolls, cumulic cryaquolls, and aquic cryoborolls (soils that have mean annual temperatures lower than 8°C at 50 cm, that are continuously or periodically saturated, and that have loamy surface horizons with >2.5 percent organic carbon). Differences in soil morphology associated with increasing wetness parallel those for other hydrosequences (Johnston et al. 1995; Castelli et al. 2000). They include increasing thickness of O and A horizons, disappearance of B_w horizons, and decreasing depth to redoximorphic features. Differences in physical and chemical properties of soils along the hydrologic gradient are due to moisture, parent material, and the interactions between these two variables. Sites with higher water tables have higher organic matter, total nitrogen, cation exchange capacity, and extractable potassium, but lower pH. Parent materials influence both physical and chemical properties of soils. Watersheds with chert, quartzite, and limestone have higher silt and clay, neutral pH, and high levels of extractable phosphorus. In contrast, watersheds characterized by acidic volcanic tuffs, rhyolites, and breccia have coarser-textured soils, low pH, and extractable phosphorus. Because parent material can influence substrate characteristics, it affects soil and water chemistry (see chapter 6) and nutrients available to plants.

Because groundwater regimes influence soil water and oxygen availability, they also affect chemical reactions and biotic processes in riparian meadow complexes. The relationships between vegetation types, groundwater depths, soil redox potentials, and soil temperature were examined for two riparian meadow complexes in the central Great Basin (Castelli et al. 2000). Groundwater depth, redox potential, and soil temperature were all strongly related to elevation and vegetation type, but there were significant differences in these relationships between sites and over the growing season (fig. 7.4). Precipitation during the study year (1998) was about 30 percent higher than the long-term average, with more spring and summer moisture. Consequently, seasonal declines were probably less than in an average or dry year for both water tables (see Big Creek data in chapter 5) and redox potentials. As for other higher-elevation hydrosequences, temperature regimes varied inversely with soil moisture regimes (Klickoff 1965). Redox values ranged from a low of about −300 millivolts at 30 centimeters for the wettest and most reducing sites (*C. nebrascensis* meadow) to a high of about +500 millivolts for the driest sites (*A. tridentata tridentata/L. cinereus* type). These values are typical of hydrosequences in the Sierra Nevada (Svejcar et al. 1992) and elsewhere (Johnston et al. 1995).

Ecophysiological Responses

The groundwater regimes of these meadow complexes influence plant rooting activity, physiological responses, and productivity (Martin and Chambers 2001a,b, 2002). Seasonal and yearly differences in water table depth over a three-year period largely determined the rooting activity (number of roots per square centimeter) and depth of mesic meadow vegetation in central Great Basin meadow complexes (fig. 7.5) (Martin and Chambers 2002). Depth to water table ranged from about 0 to 40 centimeters after peak runoff in early June to about 50 to 100 centimeters in August. Little rooting activity occurred within or at the surface of the water table, and rooting activity increased as water table elevation declined during the growing season. Similarly, for a *C. nebrascensis*–dominated site in the Sierra Nevada, minimal rooting activity was observed within the water table (Svejcar and Trent 1995).

Plant and soil water relations, photosynthesis, and biomass also are related to spatial and temporal differences in water tables. For *C. nebrascensis*, *J. balticus*, and *D. cespitosa* in a Sierra Nevada meadow, photosynthesis rates were 12.5 percent higher at streamside locations than 20 meters from the stream, where water tables were 40 centimeters deeper (Svejcar and Riegel 1998). In the central Great Basin, standing-crop biomass was influenced by grazing and restoration treatments, but the underlying controls were depth to water table and soil water content of the surface 50 centimeters. When the data were examined across sites and treatments, biomass was lowest in 1996, a low water table year, and highest in 1998, a high water table year (Martin and Chambers 2001a).

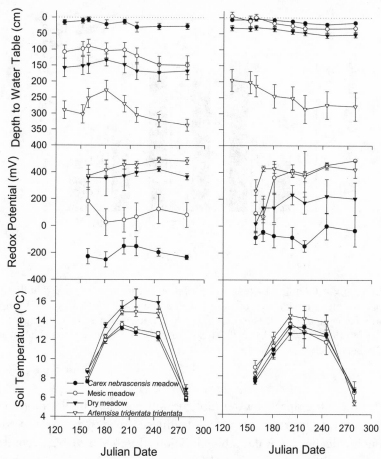

FIGURE 7.4. Depth to water table, redox potential, and soil temperature during the 1998 growing season for common vegetation types in two central Great Basin meadow complexes (modified from Castelli et al. 2000).

Indicators of Groundwater Status

Because species composition of plants is closely related to particular water table regimes, it can serve as an indicator of groundwater status in riparian ecosystems. In arid and semiarid riparian areas, few data are available on the hydrologic requirements of individual species or communities of riparian plants (but see Stromberg et al. 1996). Most of the data that do exist are for riparian trees, especially *Populus* spp. (e.g., Amlin and Rood 2002). Data for the streamside vegetation types described above allow for examination of the relative water requirements of riparian species that occur both in meadows and adjacent to streams in Great Basin watersheds. Along effluent or hydrologically losing streams that typify semiarid areas, the

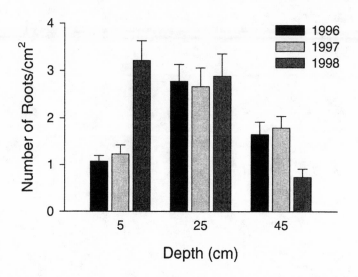

FIGURE 7.5. Rooting activity (number of roots per square centimeter) at three depths for mesic meadow vegetation in the central Great Basin. Data are from Corral Canyon; $n = 4$. Measurements were taken in August and include relatively dry years (1996, 1997) and a relatively wet year (1998) (modified from Martin and Chambers 2002).

elevation of the riparian water table usually is similar in elevation to the adjacent river or stream stage, and declines in river stage result in corresponding declines in water table elevation (Busch et al. 1992; Stromberg and Patten 1996). In the central Great Basin, total stream flow is highly variable among and within years, but peak flows are runoff dependent and consistently occur in the late spring (May–early June), after which flows decline rapidly. In these low flow systems, water depth in the thalweg (a line connecting the deepest part of the channel) at base flow is seldom greater than 10 centimeters. Thus, height above the channel may slightly underestimate depth to groundwater, but it appears to be a reasonable proxy measurement during most of the year.

The dominant species of the different streamside vegetation types exhibit individualistic but overlapping responses in terms of height above the channel bed and relative water requirements (fig. 7.6). Graminoid and forb species classified as obligate or facultative wetland species, *C. nebrascensis*, *D. cespitosa*, and *Epilobium ciliatum*, do not occur at terrace heights greater than 1 meter, and the graminoids have low occurrence rates on terraces higher than 0.25 meter. Obligate wetland shrubs, *S. exigua* and S.

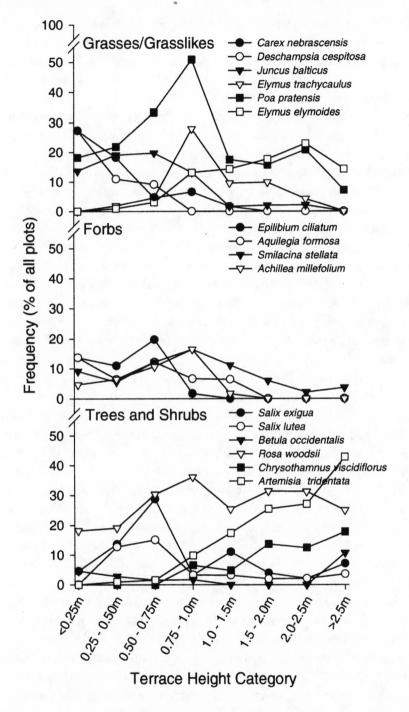

FIGURE 7.6. The frequency of the dominant species by terrace height category for the gaged drainage basins listed in table 7.1.

lutea, have high occurrence rates on terraces 0.25 to 1.0 meter high but are uncommon on terraces higher than 1.0 meter. Facultative wetland species, such as the widespread *Poa pratensis pratensis* and disturbance-adapted *R. woodsii ultramontana,* occur on all terraces, but their rates of occurrence are highest on intermediate terraces. *Juncus balticus, Aquilegia formosa,* and *Achillia millefolium* occur primarily on terraces less than 1.5 meters high, while *Elymus trachycaulus* and *Smilacina stellata* occur on terraces less than 2.5 meters high. Species that are not wetland associated, such as *Elymus elymoides, Artemisia tridentata tridentata,* and *Chrysothamnus viscidiflorus,* occur primarily on the highest terraces. Earlier studies indicate that wetland obligate species decline rapidly when terrace heights, a surrogate for groundwater depths, exceed about 0.30 meter (Stromberg et al. 1996; Castelli et al. 2000). *Carex nebrascensis,* a dominant, wetland obligate, consistently occurs at high water tables (0.30 meter in meadows; 0.38 meter on stream terraces) and appears to be a dependable indicator of average depth to water table. In contrast, large temporal and spatial variability in water table depths for species associated with higher stream terraces suggest that these species indicate only broad ranges in water table depths.

For both streamside and meadow vegetation in the central Great Basin, the variability in water table depth increases with increasing depth to water table (fig. 7.3; Castelli et al. 2000). Similar relationships between water table depth and variability have been observed in other semiarid riparian ecosystems (Stromberg et al. 1996) and elsewhere in the central Great Basin (see chapter 5). Integrated environmental variables (range in water table depth, number of days the water table was less than 30 and 70 centimeters, number of degree-days of anaerobiosis during the growing season) that incorporate the variability in water table depths over the growing season have closer relationships to the *C. nebrascensis* and mesic meadow vegetation types than water table alone (Castelli et al. 2000). Other studies have identified elevation above the stream channel and hydrologic variables as those most closely related to plant distributions (Allen-Diaz 1991; Stromberg et al. 1996). In central Great Basin meadows, integrated environmental variables were more sensitive to the spatial and temporal differences in water tables than individual species or vegetation types. They would be expected to respond more rapidly to changes in local hydrology than plant species and are probably more reliable indicators of both current water-table status and potential vegetation following stream incision.

Effects of Livestock Grazing

Montane riparian meadows in the central Great Basin have been extensively used for livestock grazing since the late 1800s. The effects of livestock grazing on riparian vegetation in the western United States are controversial and are reviewed in Kauffman and Krueger (1984), Clary and Webster (1989), Skovlin (1984), Fleischner (1994), Ohmart (1996), Belsky et al. (1999), and elsewhere. In general, livestock grazing influences riparian vegetation by (1) removing plant biomass, which allows soil temperatures to rise and results in increased evaporation, (2) damaging plants by rubbing, trampling, grazing or browsing them, (3) altering nutrient dynamics by depositing nitrogen in excreta from animals and removing foliage, and (4) compacting soil, which increases runoff and decreases water availability to plants. These effects can cause changes in plant physiology, population dynamics, and community attributes such as cover, biomass, composition, and structure. Relatively few studies exist on the effects of livestock grazing on montane meadows, and methodologies often differ, making it difficult to draw overarching conclusions about grazing effects.

BIOMASS REMOVAL

In general, moderate to heavy clipping or grazing of montane meadows can significantly decrease plant growth (Clary and Kinney 2002) and alter species composition (Green and Kauffman 1995). Native sedges (*Carex* species) and bunch grasses (e.g., *D. cespitosa*) often decline in abundance, while the widely naturalized *P. pratensis* and exotic species increase under heavy grazing (Kauffman et al. 1983; Schultz and Leininger 1990; Green and Kauffman 1995; Martin and Chambers 2001a). Species richness is often higher in grazed than not grazed plots, primarily due to the presence of nonnative grasses (e.g., *Bromus mollis, Phleum pratense*) (Green and Kauffman 1995) and low-growing forbs (e.g., *Aster occidentalis, Stellaria longipes, Taraxacum officinale*) (Martin and Chambers 2001a).

NITROGEN DEPOSITION

Nitrogen is deposited in the excreta of grazing animals and has the potential to enrich meadow soils. In Idaho, for example, autumn application of manure and urea at levels similar to those produced by cows in high-elevation sedge meadows increased standing crop biomass in the following year by almost 10 percent (Clary 1995). To assess the effects of nitrogen addition on Great Basin meadows, a slow-release nitrogen fertilizer, sulfur-coated urea (36-0-0), was applied to three mesic meadow sites at a rate of 100 kilograms per hectare in autumn 1995, 1996, and 1997, and the eco-physiological responses and community dynamics were quantified (Martin and Chambers 2001a, 2002). The type of fertilizer and application rate used in the study increases above-ground production and disease resistance in both *C. nebrascensis* and *Poa pratensis* in agricultural settings (Davis and Dernoeden 1991; Thompson and Clark 1993; Reece et al. 1994). In general, addition of nitrogen to Great Basin mesic meadows decreased rooting activity (number of roots per square centimeters) and resulted in less-negative water potentials for both *C. nebrascensis* and *P. pratensis* (Martin and Chambers 2002). Photosynthetic rates were higher than for nontreated plots early in the growing season but lower later in the season. The effects of added nitrogen on both water relations and photosynthesis could be attributed to accelerated plant phenology and earlier senescence in the nitrogen-enhanced plots. Nitrogen addition in combination with clipping to a stubble height of 5 centimeters increased biomass during all three years of the study (Martin and Chambers 2001a). The nitrogen application level used in this study probably exceeded the level of nitrogen addition due to livestock grazing in these types of meadows. Much of the nitrogen deposited by grazing animals may be lost due to volatilization and leaching (Woodmansee 1978), and vegetation removal by grazing animals may result in an export of nitrogen from grazed systems (Berendse et al. 1992). However, the results of this study indicated that relatively high levels of nitrogen addition can decrease rooting activity in mesic meadows. In meadow systems subject to high annual as well as seasonal variability, this may have long-term negative effects on species responses and community productivity.

Soil Compaction

The hoof action of grazing animals can decrease soil macropore space resulting in soil compaction. Soil compaction decreases water infiltration and leads to reduced root growth and overall lower primary productivity (Laycock and Conrad 1967; Bohn and Buckhouse 1985). In the central Great Basin, mesic and dry meadows frequently have layers of compacted soil at depths of 15 to 30 centimeters, and soil compaction has been used as an indicator of ecological condition in these systems (Weixelman et al. 1996, 1997). The effects of a one-time aeration treatment on plant ecophysiological responses and community dynamics were evaluated for three central Great Basin mesic meadows with compacted soils (Martin and

Chambers 2001a, 2002). The treatment consisted of using a 2-centimeter-diameter drill to create 30-centimeter-deep holes that were uniformly spaced 20 centimeters apart. In general, aeration increased rooting depth and activity (number of roots per square centimeters) where there was no confounding effect of water table. Also, both predawn and midday water potentials of *C. nebrascensis* and predawn water potentials of *P. pratensis* were less negative in aerated plots. The one-time aeration treatment had no effect on total standing crop biomass, but repeated treatments may have greater effects.

Interacting Effects of Hydrologic Regimes and Livestock Grazing

The outcomes of species interactions in riparian meadows are often attributed largely to anthropogenic disturbances like livestock disturbance. Based on research in the central Great Basin, it was hypothesized that water table is the primary variable influencing species responses and interactions within *C. nebrascensis* and mesic meadows in the central Great Basin but that the direct and indirect effects of livestock grazing can modify those responses and interactions. This hypothesis was examined for two widespread riparian species (*C. nebrascensis* and *P. pratensis*) that co-occur but are most abundant at different water table depths (Martin and Chambers 2001b). *Carex nebrascensis* is a sedge that has loosely arranged and widely spread tillers, while *P. pratensis* is a grass that has closely spaced and compact tillers (termed "guerilla" and "phalanx" plant architecture, respectively, by Lovett Doust 1981). Individuals of both species were grown at mid- and low water tables with or without neighbors and were either clipped or not clipped at the end of the first growing season. Water table depths measured during the growing season (May through August) varied among years. For the meadow with the most continuous record, water table depths were –32 centimeters in year 1 and –7 centimeters in year 2 for the mid-water table plots, and –69 centimeters in year 1 and –31 centimeters in year 2 for the low-water table plots. Water table depth had no effect on tillering or biomass of *C. nebrascensis*, indicating that the species is adapted to the range of water table depths examined. In contrast, growth and tillering of *P. pratensis*, a facultative upland species, was severely restricted at shallower water table depths. Clipping had little effect, possibly because clipping closer than the 10-centimeter stubble height used in the study is required to reduce tiller number and shoot mass of these species (Ratliff and Westfall 1987; Thompson and Clark 1993). *Poa pratensis* responds more rapidly than *C. nebrascensis* to disturbances that remove neighbors and create open patches. Neighborhood removal resulted in a three- to ten-fold increase in tillering for *C. nebrascensis*, but a 6- to 100-fold increase in tillering for *P. pratensis*. Comparisons of single- and mixed-species plots showed that given water table conditions favorable to both species, *P. pratensis* limited tiller production of *C. nebrascensis*.

The distribution and relative abundance of *C. nebrascensis* and *P. pratensis* are undoubtedly influenced by both the timing and duration of soil saturation and soil water availability during the growing season. Disturbances resulting from livestock grazing or other land uses that increase space and resources within the plant community allow both species to expand locally. The interactions between the two species do not seem to be related to plant architecture and can be attributed to generally greater growth rates and increased competitive ability for *P. pratensis* at lower water table depths. Livestock grazing may alter the relative competitive ability of the two species in favor of *P. pratensis*. Research on riparian meadows in the central Great Basin and elsewhere indicates that *P. pratensis* generally increases in cover in response to both grazing and clipping but decreases in cover or does not change in the absence of grazing (Kauffman et al. 1983; Schultz and Leininger 1990; Green and Kauffman 1995; Martin and Chambers 2001a). In contrast, *C. nebrascensis* shoot growth and cover increases in response to release from grazing and either remains unchanged or decreases under grazing (Ratliff and Westfall 1987; Martin and Chambers 2001a).

Hydrologic Variability and Study Designs

The high spatial and temporal variability of the water table and its importance to riparian meadow species and vegetation communities indicates that studies designed to evaluate the structure or function of these systems need to consider the water table regime (Martin and Chambers 2001a). The variability in the water table regime is often sufficient to obscure treatment effects and may explain, in part, a lack of consistent responses in both grazing and restoration studies (Clary 1995). In most cases, water table should be treated either as a main factor or as a covariate when evaluating treatment effects on plant response variables. Depending on the study design, a high number of blocks or relatively large blocks with multiple, randomly located treated plots should be used to adequately account for the high spatial variability in water tables. Finally, sampling should be conducted over relatively long time periods (three to five years) to account for the annual and seasonal variability in these systems.

Management and Restoration Implications

Results of the research conducted on upland watersheds in the central Great Basin have implications for the restoration and management of both streamside vegetation and riparian meadow complexes. Differences in basin geology and morphometry have significant effects on watershed sensitivity to both natural and anthropogenic disturbances (Swanson et al. 1998; chapter 4). In the central Great Basin, the composition and pattern of riparian vegetation is determined by the hydrogeomorphic characteristics of individual watersheds and is closely related to basin sensitivity to disturbance as indicated by past stream incision. Side-valley alluvial fans influence the geomorphic and hydrologic characteristics and vegetation patterns of riparian corridors even after stream incision. The overriding effects of the geomorphic characteristics of the watersheds on basin sensitivity to disturbance and riparian vegetation composition and pattern illustrate the importance of developing management and restoration schemes that address larger scales and incorporate hydrogeomorphic attributes.

Many of the management guidelines and restoration approaches that have been developed for riparian vegetation are based on a limited number of geomorphic and hydrologic variables, and focus primarily at stream reach scales (Goodwin et al. 1997). However, fluvial geomorphic processes and landforms at watershed-to-valley-segment scales strongly influence the occurrence and community composition of riparian plant species (Harris 1988; Bendix 1999; Wasklewicz 2001). The parameters used to categorize Great Basin watersheds according to sensitivity to disturbance, including geology and morphometry, valley width and gradient, substrate characteristics, channel gradient and uniformity, channel incision and erosion, and the relative influence of side-valley alluvial fans (chapter 4), also are the major determinants of vegetation types and associations. At valley-segment-to-stream-reach scales, water availability, as indicated by surface elevation above the water surface or stream channel and by soil texture and moisture-holding capacity, is the primary control on riparian species distributions (Hughes 1990; Stromberg et al. 1996; Wasklewicz 2001). In the central Great Basin, species occurrences and vegetation types are strongly correlated with elevation above the stream channel and with bank and channel particle sizes. Examining a broader range of scales and collecting integrated geomorphic, hydrologic, and vegetation data can improve understanding of the linkages among the abiotic and biotic components across a range of scales—from watershed to stream reach (Gregory et al. 1991).

Riparian meadow complexes are one of the highest-priority ecosystems for management and restoration in the central Great Basin because they often occur in incision-dominated watersheds and are susceptible to stream incision (chapter 5). Determining appropriate management and restoration scenarios for riparian meadow complexes begins with understanding the geomorphic and hydrologic controls on stream-incision processes. Meadow complexes in fan-dominated basins (Group 3) are located primarily upstream of side-valley alluvial fans (Miller et al. 2001; chapters 4 and 5). Incision in these basins often starts at the fans because maximum values of channel gradients and shear stress occur where axial channels traverse the fans. For fans that have been breached, waves of incision migrate upstream through the meadow complexes in the form of knickpoints. In basins with pseudostable channels (Group 4), localized incision occurs as a result of upstream migration of knickpoints and often is facilitated by groundwater sapping (chapters 4 and 5). In both types of basins, land-use activities that destabilize stream channels in or adjacent to meadows, including stream diversions, road crossings, and overgrazing by livestock, should be avoided. Restoration efforts should focus on meadow complexes with relatively stable stream channels. For fan-dominated basins, it may be possible to stabilize the axial channel at the point where it crosses the fans with grade-control structures or armoring.

Determining the site-restoration potential of degraded riparian meadow complexes requires an understanding of the relationships among hydrologic regimes, soil characteristics, and riparian vegetation. Riparian meadow complexes occur along hydrologic gradients, and soil physical and chemical properties, plant physiological processes, and plant population and community dynamics are strongly influenced by water table depths (Chambers et al. 1999; Castelli et al. 2000; Martin and Chambers 2001a,b, 2002; Wright and Chambers 2002; chapter 5). Water table depths within meadow complexes are highly variable in space and time, and the variability increases with increasing depth to the water table (Castelli et al. 2000; chapter 5). Species that require the shallowest water table depths, such as the C. nebrascensis, tolerate the least variability. In degraded areas, integrated environmental variables, such as the range in water table depth and the number of degree-days of anaerobiosis during the growing season, are more sensitive to the spatial and temporal differences in water tables than individual species or vegetation types and are accurate indicators of groundwater status and potential vegetation (Castelli et al. 2000). Riparian obligate species that require relatively high and stable water tables also are fairly consistent indicators of groundwater status. In contrast, species asso-ciated with deeper water tables indicate only broad ranges in water table depths. To accurately understand the site-restoration potential, both water table depths and vegetation should be monitored over relatively long time periods (more than three years) (Martin and Chambers 2001a), and sam-ple sizes should be large enough to account for the high spatial variability.

Overgrazing by livestock and other types of anthropogenic disturbance can alter physiological responses and competitive interactions of plant species and can exacerbate the effects of changes in groundwater levels. For example, overgrazing of riparian meadows by livestock often decreases the infiltration capacity of soils via soil compaction and alters plant physi-ological processes and population and community dynamics through vegetation removal and nitrogen deposition. Proactive management of live-stock and other anthropogenic disturbances is essential for sound man-agement and successful restoration of these ecosystems and, in incision-dominated basins, can potentially lessen the effects on ongoing stream incision.

The research described in this chapter indicates that understanding the underlying relationships among geomorphic processes, hydrologic regimes, and vegetation patterns and dynamics is required for managing and restoring riparian ecosystems. Assessing these relationships over a broader range of scales than has been done in the past—watershed to stream reach—is necessary for predicting vegetational responses to stream incision and other types of disturbance and for developing appropriate management guidelines and restoration techniques.

Acknowledgments

Many thanks to David Martin, Regine Castelli, Amy Linnerooth, Pamela Wehking, Catherine Davis, Danielle Henderson, Michael Wright, Dave Weixelman, and Desi Zamudio for their valuable contributions to this body of work. The manuscript was significantly improved by review comments from Carla D'Antonio, Jonathan Friedman, David Merritt, and Karen Zamudio.

Representative species of common vegetation types discussed in this chapter.
Vegetation types and representative species correspond with
the ecological types described in Weixelman et al. (1996) for
central Nevada.

Vegetation Type	Graminoids	Forbs	Shrubs and Trees
Carex nebrascensis meadow	*Carex nebrascensis* *Deschampsia cespitosa* *Juncus balticus*	*Geum macrophyllum* *Veronica americana*	
Mesic meadow	*Poa pratensis* *Elymus trachycaulus* *Juncus balticus* *Agrostis stolonifera*	*Aster occidentalis* *Iris missouriensis* *Stellaria longipes*	
Dry meadow	*Poa secunda* *Muhlenbergia richardsonis* *Leymus triticoides*	*Potentilla gracilis* *Achillia millefolium* *Penstemon rydbergii*	
Salix spp./mesic meadow	*Poa pratensis* *Juncus balticus*	*Smilacina stellata* *Viola sororia* *Trifolium wormskjoldii*	*Salix exigua* *Salix lutea* *Rosa woodsii*
Salix spp./mesic forb	*Leymus triticoides* *Elymus trachycaulus* *Carex praegracilis*	*Smilacina stellata* *Aquilegia formosa* *Aconitum columbianum*	*Salix exigua* *Salix lutea* *Salix lasiolepis* *Rosa woodsii*
Betula occidentalis/mesic meadow	*Poa pratensis* *Agrostis stolonifera* *Carex microptera* *Elymus trachycaulus*	*Smilacina stellata* *Aquilegia formosa* *Aconitum columbianum*	*Salix exigua* *Salix lutea* *Cornus sericea* *Rosa woodsii*
Populus tremuloides/*Symphoricarpos* spp.	*Elymus trachycaulus* *Poa pratensis* *Carex rossii*	*Aquilegia formosa* *Lupinus argenteus* *Astragalus lentiginosus*	*Populus tremuloides* *Symphoricarpos oreophilus* *Rosa woodsii*
Populus spp.	*Leymus triticoides* *Bromus carinatus* *Carex microptera*	*Smilacina stellata* *Aquilegia formosa* *Lupinus argenteus*	*Populus balsamifera* *Populus angustifolia*

Artemisia tridentata tridentata/	Leymus cinereus	Lupinus argenteus	Artemisia tridentata tridentata
Leymus cinereus	Muhlenbergia richardsonis Poa secunda Leymus triticoides	Cryptantha flavoculata Astragalus lentiginosus	Chrysothamnus viscidiflorus
Artemisia tridentata tridentata/	Poa secunda Elymus	Lupinus argenteus	Artemisia tridentata tridentata
Poa secunda	lanceolatus Leymus cinereus	Allium bisceptrum Cryptantha flavoculata	Chrysothamnus viscidiflorus

LITERATURE CITED

Allen-Diaz, B. H. 1991. Water table and plant species relationships in Sierra Nevada meadows. *American Midland Naturalist* 126:30–43.

Amlin, N. M., and S. B. Rood. 2002. Comparative tolerances of riparian willows and cottonwoods to water-table decline. *Wetlands* 22:338–346.

Auble, G. T., J. M. Friedman, and M. L. Scott. 1994. Relating riparian vegetation to present and future streamflows. *Ecological Applications* 4:544–554.

Belsky, A. J., A. Matzke, and S. Uselman. 1999. Survey of livestock influences on stream and riparian ecosystems in the western United States. *Journal of Soil and Water Conservation* 54:419–431.

Bencala, K. E. 1993. A perspective on stream-catchment connections. *Journal of the North American Benthological Society* 12:44–47.

Bendix, J. 1999. Stream power influence on southern California riparian vegetation. *Journal of Vegetation Science* 10:243–252.

Bendix, J., and C. R. Hupp. 2000. Hydrological and geomorphological impacts on riparian plant communities. *Hydrological Processes* 14:2977–2990.

Berendse, F., M. J. M. Oomes, H. J. Altena, and W. Th. Eberse. 1992. Experiments on species-rich meadows in The Netherlands. *Conservation Biology* 62:59–65.

Bohn, C. C., and J. C. Buckhouse. 1985. Some responses of riparian soils to grazing management in northeastern Oregon. *Journal of Range Management* 38:378–381.

Busch, D. E., N. L. Ingraham, and S. D. Smith. 1992. Water uptake in woody riparian phreatophytes of the southwestern United States: A stable isotope study. *Ecological Applications* 2:450–459.

Castelli, R. M., J. C. Chambers, and R. J. Tausch. 2000. Soil-plant relations along a soil-water gradient in Great Basin riparian meadows. *Wetlands* 20:251–266.

Chambers, J. C., R. R. Blank, D. C. Zamudio, and R. J. Tausch. 1999. Central Nevada riparian areas: Physical and chemical properties of meadow soils. *Journal of Range Management* 52:92–99.

Chambers, J. C., and A. R. Linnerooth. 2001. Restoring riparian meadows currently dominated by *Artemisia* using alternative state concepts—the establishment component. *Applied Vegetation Science* 4:157–166.

Clary, W. P. 1995. Vegetation and soil responses to grazing simulation on riparian meadows. *Journal of Range Management* 48:18–25.

Clary, W. P., and J. W. Kinney. 2002. Streambank and vegetation response to simulated cattle grazing. *Wetlands* 22:139–148.

Clary, W. P., and B. F. Webster. 1989. *Managing grazing of riparian areas in the Inter-mountain Region.* General Technical Report, INT-263, USDA Forest Service, Intermountain Research Station. Ogden, Utah.

Davis, C. 2000. Influences of alluvial fans on local channel geomorphology, incision and vegetation dynamics in the Toiyabe, Toquima and Monitor Ranges of central Nevada. Master's thesis, University of Nevada, Reno.

Davis, D. B., and P. H. Dernoeden. 1991. Summer patch and Kentucky bluegrass quality as influenced by cultural practices. *Agronomy Journal* 83:670–677.

Dunaway, D., S. R. Swanson, J. Wendel, and W. Clary. 1994. The effect of herbaceous plant communities and soil textures on particle erosion of alluvial streams. *Geomorphology* 9:47–56.

Fleischner, T. L. 1994. Ecological costs of livestock grazing in western North America. *Conservation Biology* 8:629–644.

Furness, H. D., and C. M. Breen. 1980. The vegetation of the seasonally flooded areas of the Pongolo River floodplain. *Bothalia* 13:217–230.

Germanoski, D., and J. R. Miller. 1995. Geomorphic responses to wildfire in an arid watershed, Crow Canyon, Nevada. *Physical Geography* 16:243–256.

Germanoski, D., C. Ryder, and J. R. Miller. 2001. Spatial variation of incision and deposition within a rapidly incising upland watershed, central Nevada. Pp. xi, 41–48 in *Sediment: Monitoring, modeling, and managing.* Proceedings of the 7th Interagency Sedimentation Conference, Vol. 2. Sponsored by the Subcommittee on Sedimentation. Mar. 25–29, 2001. Reno, Nev.

Goodwin, C. N., C. P. Hawkins, and J. L. Kershner. 1997. Riparian restoration in the western United States: Overview and perspective. *Restoration Ecology* 5:4–14.

Grant, G. E., and F. J. Swanson. 1995. Morphology and processes of valley floors in mountain streams, western Cascades, Oregon. Pp. 83–100 in *Natural and Anthropogenic Influences in Fluvial Geomorphology,* edited by J. E. Costa, A. J. Miller, K. W. Potter, and P. R. Wilcock. Geophysical Monograph 89. Washington, D.C.: American Geophysical Union.

Green, D. M., and J. B. Kauffman. 1995. Succession and livestock grazing in a north-eastern Oregon riparian ecosystem. *Journal of Range Management* 48:307–313.

Gregory, S. V., F. J. Swanson, W. A. McKee, and K. W. Cummins. 1991. An ecosystem perspective of riparian zones. *BioScience* 41:540–551.

Harris, R. R. 1988. Associations between stream valley geomorphology and riparian vegetation as a basis for landscape analysis in the eastern Sierra Nevada, California, USA. *Environmental Management* 12:219–228.

Hill, M. O. 1979. *TWINSPAN: A FORTRAN program for arranging multivariate data in an ordered two-way table by classification of the individuals and attributes.* Ithaca, N.Y.: Cornell University Press, Section of Ecology and Systematics.

Huggenberger, P., E. Hoehn, R. Beschta, and W. Woessner. 1998. Abiotic aspects of channels and floodplains in riparian ecology. *Freshwater Biology* 40:407–425.

Hughes, F. M. R. 1990. The influence of flooding regimes on forest distribution and composition in the Tana River floodplain, Kenya. *Journal of Applied Ecology* 27:475–491.

———. 1997. Floodplain biogeomorphology. *Progress in Physical Geography* 21:501–529.

Hupp, C. R., and W. R. Osterkamp. 1996. Riparian vegetation and fluvial geomorphic processes. *Geomorphology* 14:277–295.

Johnston, C. A., G. Pinay, C. Arens, and R. J. Naiman. 1995. Influence of soil properties on the geochemistry of a beaver meadow hydrosequence. *Soil Science Society
of America Journal* 59:1789–1799.

Kauffman, J. B., W. C. Krueger, and M. Vavra. 1983. Effects of cattle grazing on riparian plant communities. *Journal of Range Management* 36:685–691.

Kauffman, J. B., and W. C. Krueger. 1984. Livestock impacts on riparian ecosystems
and streamside management implications . . . a review. *Journal of Range Management* 37:430–438.

Kleinfelder, D., S. Swanson, G. Norris, and W. Clary. 1992. Unconfined compressive
strength of some streambank soils with herbaceous roots. *Soil Science Society of
America Journal* 56:1920–1925.

Klickoff, L. G. 1965. Photosynthetic response to temperature and moisture stress of
three timberline meadow species. *Ecology* 46:516–517.

Korfmacher, J. L. 2001. Reach-scale relationships of vegetation and alluvial fans in
central Nevada riparian corridors. Master's thesis, University of Nevada, Reno.

Lahde, Daniel B. 2003. Relationships between low-standard roads and stream incision
in central Nevada. Master's thesis, University of Nevada, Reno.

Laycock, W. A., and P. W. Conrad. 1967. Effect of grazing on soil compaction as measured by bulk density on a high elevation cattle range. *Journal of Range Management* 20:136–140.

Lovett Doust, L. 1981. Population dynamics and local specialization of a clonal perennial (*Ranunculus repens*). 1. The dynamics of ramets in contrasting habitats. *Journal of Ecology* 69:743–755.

Manning, M. E., and W. G. Padgett. 1992. *Riparian community type classification for
the Humboldt and Toiyabe National Forests, Nevada and eastern California.*
R4-Ecol-95-01. USDA Forest Service, Intermountain Region. Ogden, Utah.

Manning, M. E., S. R. Swanson, T. Svejcar, and J. Trent. 1989. Rooting characteristics
of four intermountain meadow communities. *Journal of Range Management*
42:309–312.

Martin, D. W., and J. C. Chambers. 2001a. Restoring degraded riparian meadows:
Biomass and species responses. *Journal of Range Management* 54:284–291.

——. 2001b. Effects of water table, clipping, and species interactions on *Carex nebrascensis* and *Poa pratensis* in riparian meadows. *Wetlands* 21:422–430.

——. 2002. Restoration of riparian meadows degraded by livestock grazing: Above-
and below-ground responses. *Plant Ecology* 163:77–91.

McCune, B., and M. J. Medford. 1999. *Multivariate analysis of ecological data
(PC-ORD).* Version 4.20. Gleneden Beach, Ore.: MjM Software Design.

Merritt, D. M., and D. J. Cooper. 2000. Riparian vegetation and channel change in
response to river regulation: A comparative study of regulated and unregulated
streams in the Green River Basin, USA. *Regulated Rivers: Research and Management* 16:543–564.

Miller, J. R., D. Germanoski, K. Waltman, R. Tausch, and J. Chambers. 2001. Influence of late Holocene hillslope processes and landforms on modern channel dynamics in upland watersheds of central Nevada. *Geomorphology* 38:373–391.

Mitsch, W. J., and J. G. Gosselink. 1993. *Wetlands.* 2nd ed. New York: Van Nostrand
Reinhold.

Neale, C. M. U. 1997. Classification and mapping of riparian systems using airborne
multispectral videography. *Restoration Ecology* 5:103–112.

Ohmart, R. D. 1996. Historical and present impacts of livestock grazing on fish and wildlife resources in western riparian habitats. Pp. 245–279 in *Rangeland wildlife*, edited by P. R. Krausman. Denver: Society for Range Management.

Ratliff, R. D., and S. E. Westfall. 1987. Dry-year grazing and Nebraska sedge (*Carex nebrascensis*). *Great Basin Naturalist* 47:422–426.

Reece, P. E., J. T. Nichols, J. E. Brummer, R. K. Engel, and K. M. Eskridge. 1994. Harvest date and fertilizer effects on native and interseeded wetland meadows. *Journal of Range Management* 47:178–183.

SAS Institute. 2000. *SAS for Windows*. Version 8. Cary, N.C.: SAS Institute.

Schultz, T. T., and W. C. Leininger. 1990. Differences in riparian vegetation structure between grazed areas and exclosures. *Journal of Range Management* 43:295–298.

Scott, M. L., J. M. Friedman, and G. T. Auble. 1996. Fluvial processes and the establishment of bottomland trees. *Geomorphology* 14:327–339.

Skovlin, J. M. 1984. Impacts of grazing on wetlands and riparian habitat: A review of our knowledge. Pp. 1001–1103 in *Developing strategies for range management*. Society for Range Management. Boulder, Colo.: Westview Press.

Stanford, J. A., and J. V. Ward. 1993. An ecosystem perspective of alluvial rivers: Connectivity and the hyporheic corridor. *Journal of the American Benthological Society* 12:48–60.

Stromberg, J. C. 2001. Restoration of riparian vegetation in the south-western United States: Importance of flow regimes and fluvial dynamism. *Journal of Arid Environments* 49:17–34.

Stromberg, J. C., and D. T. Patten. 1996. Instream flow and cottonwood growth in the eastern Sierra Nevada of California, USA. *Regulated Rivers: Research and Management* 12:1–12.

Stromberg, J. C., R. Tiller, and B. Richter. 1996. Effects of groundwater decline on riparian vegetation of semiarid regions: The San Pedro, Arizona. *Ecological Applications* 6:113–131.

Svejcar, T., and G. M. Riegel. 1998. Spatial pattern of gas exchange for montane moist meadow species. *Journal of Vegetation Science* 9:85–94.

Svejcar, T. J., G. M. Riegel, S. D. Conroy, and J. D. Trent. 1992. Establishment and growth potential of riparian shrubs in the northern Sierra Nevada. Pp. 151–154 in *Proceedings: Symposium on Ecology and Management of Riparian Shrub Communities*, compiled by W. D. Clary, E. D. McArthur, D. Bedunah, and C. L. Wamboldt. General Technical Report INT-289, USDA Forest Service, Intermountain Research Station. Ogden, Utah.

Svejcar, T. J., and J. D. Trent. 1995. Gas exchange and water relations of Lemmon's willow and Nebraska sedge. *Journal of Range Management* 48:121–125.

Swanson, F. J., S. L. Johnson, S. V. Gregory, and S. A. Acker. 1998. Flood disturbance in a forested mountain landscape. *BioScience* 48:681–689.

ter Braak, C. J. F. 1986. Canonical correspondence analysis: A new eigenvector technique for multivariate direct gradient analysis. *Ecology* 67:1167–1179.

Thompson, D. J., and K. W. Clark. 1993. Effects of clipping and nitrogen fertilization on tiller development and flowering in Kentucky bluegrass. *Canadian Journal of Plant Science* 73:569–575.

Thorne, C. R. 1990. Effects of vegetation on riverbank erosion and stability. Pp. 125–144 in *Vegetation and erosion*, edited by J. B. Thorne. New York: John Wiley & Sons.

USDA NRCS (United States Department of Agriculture, Natural Resources Conservation Service). 2002. *The PLANTS Database*. Version 3.5. http://plants.usda.gov. National Plant Data Center, Baton Rouge, Louisiana

Van Splunder, I., L. A. C. J. Voesenedk, H. Coops, X. J. A. De Vries, and C. W. P. M. Blom. 1996. Morphological responses of seedlings of four species of Salicaceae to drought. *Canadian Journal of Botany* 74:1988–1995.

Wasklewicz, T. A. 2001. Riparian vegetation variability along perennial streams in central Arizona. *Physical Geography* 22:361–375.

Wehking, P. M. 2002. The role of the seedbank in the restoration of a basin big sagebrush dominated riparian ecosystem to dry meadow. Master's thesis, University of Nevada, Reno.

Weixelman, D., D. Zamudio, K. Zamudio, and K. Heise. 1996. *Ecological type identification and ecological status determination*. R4-ECOL-96-01. USDA Forest Service. Ogden, Utah.

Weixelman, D. A., D. C. Zamudio, K. A. Zamudio, and R. J. Tausch. 1997. Classifying ecological types and evaluating site degradation. *Journal of Range Management* 50:315–321.

Wolman, M. G. 1954. A method of sampling coarse river-bed material. *Transactions of the American Geophysical Union* 35:951–956.

Woodmansee, R. G. 1978. Additions and losses of nitrogen in grassland ecosystems. *BioScience* 28:448–453.

Wright, M. J., and J. C. Chambers. 2002. Restoring riparian meadows currently dominated by *Artemisia* using threshold and alternative state concepts—aboveground vegetation response. *Applied Vegetation Science* 5:237–246.

Chapter 8

Explanation, Prediction, and Maintenance of Native Species Richness and Composition

Erica Fleishman, Jason B. Dunham,
Dennis D. Murphy, and Peter F. Brussard

Explanation of the distribution of species long has been a focus of theoretical and applied ecology (Ricklefs and Schluter 1993; Rosenzweig 1995;
Scott et al. 2002). Subdisciplines including biogeography, landscape ecology, and conservation biology address where species occur, the mechanisms that help generate those patterns, and the extent to which humans
may intentionally or inadvertently influence species distributions. Studies
of past, present, and future distributions of species in the Great Basin have
yielded landmark contributions to all of these fields (e.g., McDonald and
Brown 1992; Murphy and Weiss 1992; Lawlor 1998; Grayson 2000). The
theory of insular biogeography, for example, drew heavily from research in
the montane "islands" of the Great Basin, isolated from the surrounding
"sea" of sagebrush as the regional climate became warmer and drier after
the Pleistocene (e.g., Brown 1971; Lomolino 1996). The flora and fauna
of the Great Basin also featured prominently in the earliest forecasts about
the biological effects of rising concentrations of carbon dioxide and other
greenhouse gases (e.g., McDonald and Brown 1992; Murphy and Weiss
1992). Increasing awareness of natural and anthropogenic changes in the
environment of the Great Basin has intensified efforts to obtain information that may be relevant to understanding, maintaining, and restoring native species and ecosystems.

This chapter provides an overview of four approaches to studying and
managing faunal distributions in the Great Basin: documentation of historical changes; development of explanatory and predictive models; application of surrogate species, such as indicators or umbrellas, as planning
tools; and use of island biogeographic theory to anticipate ecological effects of climate change. These approaches are illustrated using case stud-

232

ies of native fishes, butterflies, and birds. Both single-species and multiple-species approaches are considered, and the ability of each approach to provide improved guidance for management of ecosystem composition, structure, and function is examined.

It is vital not only to draw strong ecological inferences from scientific investigations but also to conduct research that is transferable in space and time and has practical applicability. Therefore, most of the work highlighted in this chapter focuses on links between target species and aspects of the physical environment, such as elevation or topographic heterogeneity. Digital spatial data are increasingly accurate and cost effective. Because similar physical variables often are related to both faunal and floral diversity patterns at multiple biological levels—from individual species to assemblages, and across a range of spatial scales—the focus of work presented here is particularly useful for synthetic research, management, and restoration initiatives like the Great Basin Ecosystem Management Project.

Historical Changes

Ecological understanding of faunal distributions and the ability to apply that knowledge in a predictive context can be improved by elucidating how species distributions shift in response to natural and anthropogenic environmental change. One clear strategy for acquiring that knowledge is to document shifts in biodiversity patterns through time and correlate those shifts with known environmental perturbations. In many cases, however, the quantity and quality of historical data are limited. Another promising tactic is to explore abiotic and biotic variables that may help explain and predict species distributions.

Data on faunal distributions prior to the postsettlement period largely have been drawn from paleoecological evidence. Although fossil records are incomplete for most faunal groups, species distributions during the mid-late Holocene sometimes can be inferred from remains in pack rat middens and from pollen deposits, which provide information on climate and the distribution of vegetation resources (chapter 2).

Some knowledge of postsettlement faunal distributions has been drawn from historical (about 1850 to 1950) and more-recent field surveys. Unfortunately, issues related to access, survey methods, and variation in species occurrence complicate efforts to assemble reliable databases on species distributions in the Great Basin. Historical presence records usually are credible, especially if the observer was a reputable naturalist, but

FIGURE 8.1. (A) Birch Canyon as photographed by members of the Linsdale expedition in the 1930s, and (B, next page) as photographed by Peter Goin in 2001. Linsdale expedition photograph courtesy of Archives, Museum of Vertebrate Zoology, University of California, Berkeley. Goin photograph courtesy of Peter Goin, University of Nevada, Reno.

lack of data rarely can be interpreted as a legitimate absence record. Much of the Great Basin is topographically complex and difficult to access by foot, let alone by vehicle. Consequently, historical survey efforts across the region have been uneven, and records from museums and the archives of resource agencies may misrepresent species distributions across the entire Great Basin. Inaccessibility continues to plague contemporary surveys. Nonetheless, basic presence data from faunal surveys throughout the twentieth century provided baseline information from which inferences can be drawn regarding historical changes in species distributions (e.g., Hubbs and Miller 1948; Brown 1971, 1978; Johnson 1978; Smith 1978; Dobkin and Wilcox 1986; Wilcox et al. 1986). Moreover, in the few cases in which historical photographs exist, repeat photography provides a novel and compelling way to "see" changes in vegetation over many decades (fig. 8.1).

A substantial body of ecological literature is devoted to survey protocols, allocation of sampling effort in relation to area and species richness, and species-specific detection probabilities. But, when present-day surveys are compared to historical records, poor documentation of methods and observations can inhibit detection of faunal change. Two problems are especially prevalent in older data sets: failure to record absences (i.e., locations

where species were not encountered) and, to a lesser extent, failure to record common species. For example, it appears that relatively ubiquitous species were not recorded during surveys of butterflies in the Toiyabe Range in the 1930s (Fleishman et al. 1997). Furthermore, it is possible that species with particular sensitivity to postsettlement human disturbance were lost from the Great Basin before any occurrence records were obtained.

Temporal variation in the distributions of many animals, especially those with short generation times, also hampers documentation of faunal changes. Most biological surveys are relatively short term, involving a few years or even a single year. Moreover, the distribution of many species varies through time in response to changes in weather, resource availability, and interspecific interactions; the average magnitude of that variation differs among taxa (Scott 1986; Pollard 1988; Fleishman et al. 1997). Thus, the point at which a "snapshot" of occupancy is taken can affect both estimation of current species richness, and occurrence and appraisal of subsequent changes. Accordingly, more temporally extensive data generally are expected to yield more accurate assessments (Hanski et al. 1996; Hanski 1999; Moilanen 2000).

Unfortunately, management decisions rarely can be delayed until detailed species records are available. Ecologists and managers, therefore, seek tools to guide land-use planning in the absence of complete information. Within an ecosystem, relationships between species distributions and major physical gradients should remain relatively consistent over time—at least from a human perspective if not from an evolutionary or geological perspective—regardless of whether species distributions actually may have shifted in response to environmental change. The process of developing and testing hypotheses about key environmental variables that affect species distributions allows one to infer how proposed management and restoration might be reflected by future species distributions (Fleishman and Mac Nally, forthcoming). If convincing data exist on past changes in ecosystem structure and function, then, to a limited extent, it also may be possible to estimate potential historical faunal patterns. Thus, predictive models of species distributions and application of surrogate species are two of the most frequently touted shortcuts for setting management objectives and developing effective strategies to achieve those targets.

Predictive Models of Species Distributions

This section addresses methods to explain and predict two complementary aspects of species distributions: species richness (the number of species in a specified location) and occurrence (presence/absence) patterns of individual species. Species richness is an essential component of biological diversity with broad relevance to management and restoration (Stohlgren et al. 1995; Oliver and Beattie 1996; Longino and Colwell 1997; Ricketts et al. 1999). Species richness also is an intuitive variable that is easily understood by diverse stakeholders. Measures of species richness, however, do not directly address species composition (that is, which species are present). Moreover, variables that influence species richness may have little effect on the distributions of individual species. Rare or vulnerable species need not occur in the locations with greatest species richness, and the distributions of species of special concern may not overlap (Cody 1986; Thomas 1995; Fagan and Kareiva 1997; Freitag et al. 1997; Rubinoff 2001). It is, therefore, important to explain and predict both species richness and occurrence (Margules and Pressey 2000).

Faunal distributions in the Great Basin, like many ecosystem-level phenomena in the region (chapters 3, 4, and 9), rarely have a stable or equilibrium state. Species that tend to occur in many small and ephemeral local populations with dynamics that are linked by limited dispersal (i.e., metapopulations; Hanski and Gilpin 1997; Hanski 1999) may require special management consideration because maintenance of suitable but temporarily unoccupied habitat is critical to their long-term persistence. The relevance of spatial and temporal scale to relationships among species richness, species occurrence, and environmental parameters is increasingly apparent (Smith et al. 1998). Explanatory or predictive models, however, may be valid only at one spatial or temporal scale—or for one location or time period (Wiens 1989; Cooper et al. 1998). Testing the extent to which such models are transferable in space and time is essential.

There is no single or ideal way to model species richness or occurrence. Instead, there are numerous complementary alternatives. Species richness or occurrence potentially can be explained as a function of variables related to the resource requirements of the assemblage or species of interest, such as food sources or percent cover of woody vegetation (Braithwaite et al. 1989; Austin et al. 1990; Lindenmayer et al. 1990; Scott et al. 2002). Obtaining such data, however, can be time consuming and expensive, particularly throughout extensive areas.

A second approach employs variables that can be quantified easily, at fine resolution and over large areas, using remote-sensing data or a geographic information system (GIS) (Busby 1991; Caicco et al. 1995; Neave and Norton 1998; Fleishman et al. 2001b). Data sets and methods for deriving these variables are increasingly available and affordable, and obtaining values for many variables does not require field visits (Austin et al. 1990; Guisan and Zimmerman 2000; Jackson et al. 2000). These are important practical advantages in the Great Basin. Species occurrence models built using remote sensing and GIS data are well suited to restoration efforts because they can be easily linked with GIS-based models of alternative revegetated landscapes (Lambeck 1997; Bennett 1999; Huxel and Hastings 1999). Connecting occurrence models with revegetation models allows one to estimate the quantity and distribution of suitable habitat for each species that would be available under alternative scenarios. Thus, one can gauge the overall potential of each alternative to achieve specified ecological objectives (e.g., Rieman et al. 2000).

A third popular alternative for explaining species richness—rarely tested empirically—is to employ one or more *indicator species*, the presence of which is correlated with the richness of its (or their) taxonomic group (Kremen 1994; Pearson 1994; Prendergast 1997; Carroll and Pearson 1998). Reliable indicators of species richness, if they can be found (Niemi et al. 1997; Scott 1998), offer several practical benefits. For example, if the indicators are easier to detect—especially by inexperienced observers—than related species, it may be considerably faster and cheaper to monitor the indicators than to conduct comprehensive surveys (Gustafsson 2000).

Species Richness: Butterflies

Predictive models of species richness of butterflies in the central Great Basin have been developed as a function of (1) topographic and climatic variables derived from GIS, and (2) the occurrence of certain indicator species (see below). From 1996 to 1999, standard methods were used to conduct comprehensive inventories of resident butterflies in forty-nine locations in ten canyons in the Toquima Range (Fleishman et al. 1998, 2000). The Toquima Range data, which included fifty-six species of butterflies, were used to build the species richness models.

For each inventory location, a GIS was used to derive fourteen predictor environmental variables that might reasonably be expected to affect, and thus to predict, butterfly distributions, including geographic coordinates, elevation, slope and aspect, area, precipitation, solar insolation, topographic exposure and heterogeneity, and distance to the nearest source of permanent water (for a complete description see Fleishman et al. 2001b; Mac Nally et al. 2003). Squares of the environmental variables also were used to accommodate potential nonlinear responses (e.g., declines in species richness at extremes of a variable).

Ordinary multiple linear regression is not ideal for modeling richness data because the error distribution of richness data is expected to be Poisson rather than Gaussian (normal) (Cameron and Trivedi 1998). Species richness (the dependent variable) was fitted by using Poisson regression. With many independent variables, Schwarz's information criterion (SIC) (Schwarz 1978) is an effective and statistically robust method to identify the most efficient model (Mac Nally 2000). Every possible permutation of predictor variables is calculated and SIC for each is computed—the minimum SIC is sought. SIC is an "optimal" statistic, a compromise between model fit (ability to explain observed variation or deviance in the dependent variable) and model complexity (number of predictor variables). In addition, hierarchical partitioning was used to identify the most likely causes of variation in species richness (Chevan and Sutherland 1991; Mac Nally 2000). Hierarchical partitioning jointly considers all possible models and is designed to alleviate problems of multicollinearity among predictor variables. The increase in model fit associated with each predictor variable is estimated by averaging its additional explanatory power in all models in which that variable appears.

The model that was obtained suggested that species richness of butterflies in the central Great Basin can be predicted using just three variables that are easy to quantify across virtually any landscape: elevation, the square of elevation, and a measure of local topographic heterogeneity. These variables were included in the minimum-SIC model and also had substantial independent explanatory power (Mac Nally et al. 2003). The model explained 57 percent of the total deviance (the Poisson-regression equivalent of variance) of observed species richness of butterflies (Mac Nally et al. 2003; fig. 8.2).

It is possible to infer why elevation and topographic heterogeneity were correlated with species richness. For example, species richness tended to increase as elevation increased, although a negative coefficient associated with the square of elevation indicated a flattening of the curve at lowest and highest elevations. Overall, the relationship between species richness and elevation probably reflects a gradient in climatic severity (Fleishman et al. 2000). In the Toquima Range, low elevations are dry, with few larval host plants and adult nectar sources. Climate also may constrain species richness at higher elevations in the Toquima Range. Temperature, precipitation, and wind conditions at high elevations often limit butterfly flight time and reproduction (Kingsolver 1983, 1989; Springer and Boggs 1986; Dennis and Shreeve 1989; Dennis 1993; Boggs and Murphy 1997). Nonetheless, because much of the Toquima Range is relatively low (below 3,000 meters), higher elevations are not extremely harsh for butterflies. Species richness of butterflies also increased with increasing topographic heterogeneity. Varied topography may support correspondingly diverse plant communities, offer numerous locations for seeking mates, and provide shelter from extreme weather events (Scott 1986).

From 2000 to 2002, the same field methods were used to conduct inventories of butterflies in a total of thirty-nine locations in eight canyons in the nearby Shoshone Mountains and to derive environmental variables for those locations. The Shoshone Mountains data will be used to test the predictions of the species richness model developed using data from the Toquima Range.

Species Occurrence: Butterflies

The same inventory data and topographic and climatic variables described in the species richness case study above were used to predict occurrence (presence or absence) of individual species of butterflies in the central Great Basin. Meaningful models were obtained for thirty-six of the fifty-six resident species of butterflies recorded from the Toquima Range (Fleishman et al. 2001b). The models explained 8–72 percent of the

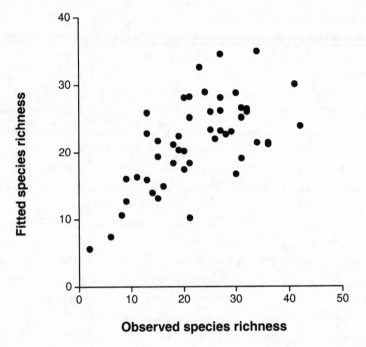

FIGURE 8.2. Fitted species richness versus observed species richness of butterflies in the Toquima Range. Fitted values are based on a model of species richness as a function of elevation, the square of elevation, and a measure of local topographic heterogeneity.

deviance in occurrence of those species (mean = 34 percent, SD = 18 percent).

Validation data collected from the Shoshone Mountains in 2000–2001 were used to assess the success of the occurrence predictions. Predictions were relatively successful overall (73 percent), with success rates for predicted absences uniformly higher than for predicted presences. Increasing the temporal extent of data from one to two years elevated success rates for predicted presences but decreased success rates for predicted absences, leaving overall success rates essentially the same. The latter result is not surprising because species composition in almost every ecological system is temporally variable to some extent. On one hand, it is possible that a species will be present at a study location in some years but absent in others, perhaps because weather conditions are unfavorable; thus, over time, the success of presence predictions should increase. On the other hand, many species eventually will appear as "accidentals" in locations that are outside their typical distributional range. Thus, the success of absence predictions is likely to decrease somewhat over time.

Although species occurrence rates (proportion of locations in which each species was found) were correlated between the modeling and validation data sets (Spearman's $r_s = 0.56$, $P < 0.001$), occurrence rates for many species increased or decreased substantially, and erroneous predictions were more likely for those taxa (fig. 8.3). Model fit (measured by the proportion of explained deviance in the explanatory model) was an indicator of the probable success rate of predicted presences ($r_s = 0.59$, $P < 0.01$) but not of predicted absences or overall success rates. The difference in occurrence may be a temporal effect (different sets of years), a geographic effect (Toquima Range versus Shoshone Mountains), or both. Ongoing work will allow for discrimination among these effects, improving both ecological understanding and predictive capacity.

Species Occurrence: Fishes

Native fishes in the Great Basin fall into two general categories with respect to their patterns of distribution: isolated endemics, and widespread species that may be locally rare (Minckley and Deacon 1991). For isolated endemic species, the conservation strategy is relatively simple in concept: maintain or restore existing habitat, and consider translocations to provide insurance against local extirpations. For widespread species, developing an appropriate conservation strategy can be far more difficult, because managers must contend with a large number of local populations and potentially suitable patches of habitat. Time and money always are limited and typically can be allocated only to a small proportion of populations or habitat patches.

Much recent work in the Great Basin has focused on Lahontan cutthroat trout (*Oncorhynchus clarki henshawi*), a widespread taxon listed as threatened under the U.S. Endangered Species Act. The distribution of Lahontan cutthroat trout in the western Great Basin is related to both local and regional environmental gradients. Cutthroat trout in this region occur almost exclusively in small streams ranging from less than 1 meter to 6 meters in width during summer low flows. Within occupied streams, the downstream distribution limit of cutthroat trout is related to three factors: perennial surface flow, maximum water temperature during the summer, and the presence of nonnative trout (Dunham et al. 1999, 2000, 2003). Among streams, the elevation of the downstream distribution limits of cutthroat trout can be predicted by summer-air-temperature gradients. At the within-stream scale, however, summer air temperatures do not vary enough to be useful for predicting fish distributions.

Although thermal gradients have predictable effects on the distribution

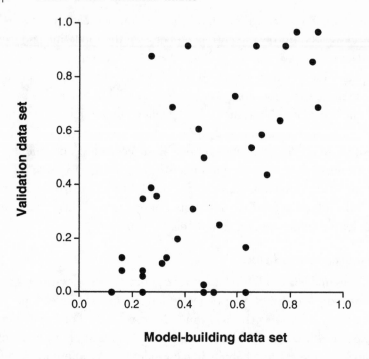

Model-building data set

FIGURE 8.3. Species occurrence rates (proportion of inventory locations in which each species was found) in the model-building (Toquima Range) versus validation (Shoshone Mountains) data sets (Spearman's $r_s = 0.56$, $P < 0.001$).

of cutthroat trout, the effect of nonnative trout on cutthroat trout appears to be less predictable. In some localities, cutthroat trout co-occur with nonnative trout, but in other localities the distribution of cutthroat trout is dramatically decreased when nonnatives are present. Nonnative trout are believed to be a major cause of declines of cutthroat trout, but no single mechanism to explain their impact has been identified (Young 1995; Dunham et al. 2002a).

Watersheds that provide thermal conditions suitable for cutthroat trout exist throughout the western Great Basin, but local populations are most likely to occur in relatively large, interconnected stream-reach (patch) complexes (Dunham et al. 1997, 2002b). Most stream reaches with suitable thermal conditions are quite small or isolated and do not currently support local populations. If the goal is to facilitate recovery of Lahontan cutthroat trout, then restoration of stream flows, restoration of suitable thermal conditions, and eradication of nonnative trout is especially important in smaller streams. Research on cutthroat trout demonstrates that the type and magnitude of threats to the persistence of native species can vary among localities. Information on these threats may be useful for spatially explicit prioritization of management actions (e.g., Rieman et al. 2000; Dunham et al. 2002a).

Surrogate Species

Efforts to predict species richness and occurrence suggest that relatively simple models can be used effectively to understand and predict contemporary faunal distribution patterns. Data used to build and validate the models described above were collected using rigorous, standardized protocols implemented over many locations and years. Unfortunately, with limited time and money, it will be difficult to conduct efforts of similar magnitude for the full diversity of faunal assemblages present in the Great Basin. Even relatively well-known taxonomic groups have not yet been surveyed comprehensively across the region. For example, since the early 1990s, researchers have detected many populations that earlier workers did not observe, including mammals in select mountain ranges (Grayson and Livingston 1993; Lawlor 1998; Grayson 2000; Grayson and Madson 2000) and native fishes in isolated stream systems (Hepworth et al. 1997). For lesser-known faunal groups, many species are still being described (Weaver and Myers 1998; Hershler 1998, 1999; Kulkoyluoglu 2000; Christopher and Fugate 2001; Sada et al. 2001). It is unlikely that exhaustive inventory data will be available for most taxonomic groups in the Great Basin within the foreseeable future. A popular suggestion for addressing this dilemma is to use surrogate species.

In theory, some species can serve as reliable and cost-effective measures of other variables that are difficult and expensive to measure directly, including total species richness; ecosystem functions, such as primary productivity, rates of nutrient cycling, and water flows; or ecosystem "integrity" (e.g., Franklin 1988; Noss 1990; Angermeier and Karr 1994; Lindenmayer et al. 2000). This reasoning has led to a variety of *surrogate species* concepts (Caro and O'Doherty 1999), including umbrella species (Andelman and Fagan 2000; Fleishman et al. 2000), indicator species (Landres et al. 1988; Noss 1990; Landres 1992; Lindenmayer et al. 2000), keystone species (Mills et al. 1993; Fauth 1999), ecosystem engineer species (Jones et al. 1994; Jones 1997; Coleman and Williams 2002; Reichman and Seabloom 2002), flagship species (Leader-Williams and Dublin 2000), and focal species (Lambeck 1997; Zacharias and Roff 2001; table 8.1).

Experimental evidence to validate the utility of surrogate species is sparse. This is not surprising, because the lack of resources for inventories, monitoring, and research is a primary motivation for using surrogate species. Furthermore, guidance on how to select effective surrogate species is lacking, and there have been few efforts to identify specific management scenarios in which surrogate species are most likely to be useful. As the fol-

TABLE 8.1.

Definitions of different categories of surrogate species

Category	Definition
Umbrella	Species whose conservation confers a protective umbrella to numerous co-occurring species
Indicator	Species whose distribution, abundance, or population dynamics can serve as substitute measures of the status of other species or environmental attributes
Keystone	Species that significantly affects one or more key ecological processes or elements to an extent that greatly exceeds what would be predicted from its abundance or biomass
Ecosystem engineer	Species that, via morphology or behavior, modifies, maintains, and creates habitat for itself and other organisms
Flagship	Charismatic species that serves as a symbol to generate conservation awareness and action
Focal species	Species used, for any reason, to help understand, manage, or conserve ecosystem composition, structure, or function

lowing case studies from the Great Basin illustrate, workers have begun to address these issues by developing objective methods for selection of umbrella and indicator species.

Umbrella Species

The concept of umbrella species—species whose conservation might confer a protective "umbrella" to numerous other species—is straightforward and appealing. It is often faster and cheaper to sample a few species than to inventory an entire assemblage. Therefore, umbrella species should reduce the time and money that must be invested in collecting data to prioritize land-use alternatives.

Selection of umbrella species ought to be prospective; in practice, it almost always has been retrospective. Species typically have been suggested as umbrellas not on the basis of their geographic distribution or life history, but because they are legally protected. As a result, conservation biologists and land managers have been restricted to asking—after the fact—whether additional species will benefit from the conservation of listed species.

Lahontan cutthroat trout, for example, are listed as threatened under the U.S. Endangered Species Act and are an element of aquatic ecosystems throughout much of the Great Basin. However, conservation of this species is not likely to be an effective mechanism for conserving many additional aquatic or amphibian species. Other native fishes in the Great Basin frequently have resource requirements or geographic ranges different from those of cutthroat trout. Moreover, many aquatic invertebrates (e.g., spring snails, ostracods, caddisflies) are poorly known and often inhabit streams and springs that do not support fishes of any kind (Sada et al. 2001). In the face of limited information on species' ecology and distributions, the best strategy for conserving the diversity of aquatic species native to the Great Basin may be to identify and protect representatives of a wide range of aquatic cover types (Palik et al. 2000) and to maintain known hydrological connections and distinct hydrological units whenever possible (Angermeier and Winston 1999; Sada et al. 2001).

On a more positive note, recent work on butterflies and birds suggests that effective umbrella species can be selected using objective ecological criteria (Fleishman et al. 2000, 2001c). Ecologists have identified three key aspects of a species' distribution and biology—co-occurrence of species, occurrence rate, and sensitivity to human disturbance—that should be considered when selecting umbrella species. These three factors were then used to develop a numerical index that measures the potential of each species to serve as an umbrella for other members of its regional taxonomic group (Fleishman et al. 2000).

Perhaps the most important criterion for selecting umbrella species is co-occurrence—the proportion of species in the same group of animals or plants that is present where a potential umbrella species occurs. For example, an average of 78 percent of the butterfly species recorded from canyons in the Toiyabe Range occurred in the canyons in which the Apache silverspot butterfly (*Speyeria nokomis apacheana*) was present. Co-occurrence has rarely been considered explicitly in attempts to identify umbrella species. Instead, it often has been assumed that protecting species with large home ranges will conserve resources for species that have smaller home ranges. But species with large home ranges are often habitat generalists, and it may be impossible, and unnecessary, to protect all locations where they occur. Also, because species richness tends to vary considerably over small distances, conservation of only a portion of the area occupied by a species with a large home range may miss locations with the greatest species richness. Therefore, selection of umbrella species should focus not on how many places a species occurs, but rather on how species-rich those places are. To illustrate, the Apache silverspot butterfly was present in just seven of the canyons that were surveyed. However, those seven canyons had greater average species richness of butterflies than did the eighteen canyons occupied by the swallowtail butterfly *Papilio rutulus*.

Occurrence rate is the second criterion for selecting umbrella species. An ideal umbrella species is neither ubiquitous nor extremely rare, but instead falls between those two extremes. Ubiquitous species are unlikely to serve as effective umbrellas because they inevitably occur in many locations with relatively low species richness, and it is not feasible to protect all areas in which they are present. Rare species, too, are unlikely to serve as effective umbrellas because they occur in so few locations.

The third consideration for selecting umbrella species is sensitivity to human disturbance. From a management perspective, it tends to be more useful to characterize sensitivity to human land use than sensitivity to natural phenomena, such as weather extremes; it is presumed that sensitive species will provide a protective umbrella for other species that are equally or less sensitive to human activities. Different species may respond quite differently to similar disturbances, and even the same species may respond differently to different disturbances. Therefore, the index of sensitivity can be tailored to any taxonomic group or ecosystem (Fleishman et al. 2001c). For example, urbanization has considerable adverse influence on native birds in broadleaf forests in the eastern United States. Parameters such as nest height and territory size can be used to assess the sensitivity of birds to urbanization (Brown 1985; Ehrlich et al. 1988; Hansen and Urban 1992). By contrast, one of the primary human disturbances affecting butterflies in broadleaf forests is forest fragmentation. Sensitivity of butterflies to forest fragmentation can be assessed using parameters such as larval host plant specificity and home range size (Garth and Tilden 1986; Scott 1986; Iftner et al. 1992; Loder et al. 1998).

Research suggests that umbrella species chosen with co-occurrence of species, occurrence rate, and sensitivity to human disturbance in mind may be an efficient way to help achieve a target level of species protection (Fleishman et al. 2001c). Data on bird and butterfly species from three different ecosystems—chaparral, broadleaf forest, and sagebrush steppe—were used to test whether objectively selected "umbrella species" were more effective as umbrellas than randomly selected species. Results suggested that umbrella species could be used to identify a smaller subset of locations for conservation while still protecting the same proportion of species. In other words, umbrella species can help choose where to locate competing land uses when the amount of land dedicated to conservation is limited.

Using an umbrella species drawn from one taxonomic group is not likely to help protect a large proportion of the species in other taxonomic groups. For instance, it may be possible to identify several species of birds whose conservation is likely to confer a protective umbrella to numerous co-occurring species of birds, but birds are unlikely to function as effective umbrellas for other taxonomic groups (also see Andelman and Fagan 2000; Rubinoff 2001). For butterflies and birds, umbrella species identified using objective methods were no more effective than randomly selected species for cross-taxonomic applications (Fleishman et al. 2001c; fig. 8.4).

Indicator Species

"Indicators" of species richness have distributions that are correlated with species richness of their taxonomic group (Kremen 1994; Pearson 1994; Prendergast 1997; Carroll and Pearson 1998). For butterflies in the Great Basin, it seemed that widespread species would not be useful for modeling variation in species richness (and so, would have little potential to serve as indicator species), while restricted species, which occur at relatively few sites, often have highly specific ecological requirements that are not shared with many other species. Therefore, of the fifty-six resident species of butterflies recorded from the Toquima Range (see above), only the twenty-two species occurring in more than 30 percent and less than 70 percent of the forty-nine study locations were considered as potential indicator species. Thus, species richness at each location, including the potential indicator species, was modeled as a function of the incidence of a set of indicator species drawn from those twenty-two species.

The modeling process identified a set of five indicator species whose incidences accounted for 88 percent of the deviance in butterfly species richness in the Toquima Range (Mac Nally and Fleishman 2002): *Ochlodes sylvanoides, Everes amyntula, Euphilotes ancilla, Speyeria zerene,* and *Coenonympha tullia.* These five indicator species encapsulate a diversity of life-history characteristics found among the resident butterfly fauna of the biogeographic region. This may explain why this particular suite of species was so strongly associated with variation in species richness. For example, the phenologies of flight activity of the species span the field season. In addition, the group of indicator species includes taxa with varied larval host plants. Two species feed on different genera in the family Poaceae and one each feeds on families Polygonaceae, Fabaceae, and Violaceae.

Predictions from the model based on these five indicator species were strongly correlated with observed values of butterfly species richness in twenty-nine sites in the Shoshone Mountains (Pearson rank-correlation 0.799, fig. 8.5). More than 90 percent of the observed species richness values fell within the 95 percent credible intervals of the predictions. Although the number of indicator species contributed to overall species richness in the validation locations (maximum difference of four), this explained little of the difference between locations with the fewest and the most species (thirty-two species). The average absolute deviation was 3.6 species, but

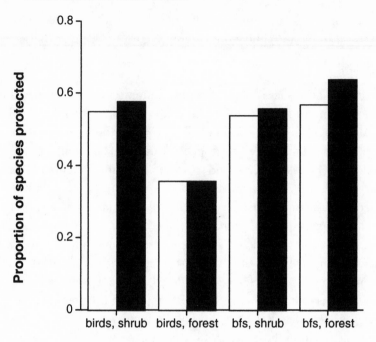

FIGURE 8.4. Proportion of species in a given taxonomic group and ecosystem that would be protected using umbrella species from a different taxonomic group (open bars) versus species drawn at random from the same taxonomic group (black bars). Bfs = butterflies; shrub = coastal chaparral shrubland in California; and forest = broadleaf forest in Ohio.

four sites contributed disproportionately to this deviation. The latter sites had from seven to fourteen fewer species than predicted, possibly because a drought cycle during the inventory period eliminated otherwise reliable sources of running or standing water in those sites. The average absolute deviation for the other twenty-five sites was just 2.7 species. A demonstrably effective model of species richness as a function of indicator species is one of several tools that may help to produce increasingly well-informed strategies for addressing diverse management objectives. On a cautionary note, however, identification of appropriate and effective surrogate species frequently requires considerable research and validation.

Anticipating Future Distributions of Species

Desert ecosystems are thought to be highly responsive to environmental changes, including shifts in temperature and precipitation, invasion by non-native plants, and altered disturbance regimes (Sala et al. 2000; Smith et al.

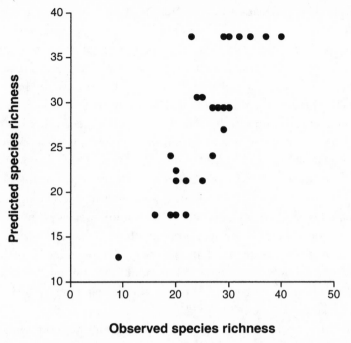

FIGURE 8.5. Mean predicted species richness versus observed species richness values for all model validation sites in the Shoshone Mountains. Predictions were obtained using a model of species richness as a function of five "indicator" species.

2000). For example, declines in species richness and changes in species composition of native plants and animals in the Great Basin are anticipated if recent climate change predictions—2–3°C increases in temperature, a 10 percent decrease in summer precipitation, and a 15–40 percent increase in precipitation during other seasons (U.S. EPA 1999)—prove accurate (McDonald and Brown 1992; Murphy and Weiss 1992; Grayson 2000; Fleishman et al. 2001a). Active partnerships among interdisciplinary teams of researchers and land mangers help to develop understanding of the effects of land use and climate change in the Great Basin, and, in turn, to develop practical alternatives for maintaining and restoring sustainable ecosystems.

Island biogeography has frequently been invoked to explain faunal distributions in montane ecosystems, in which many species have been isolated by elevation (Wilcox 1980; Myers 1986; Meffe and Carroll 1994; Guisan et al. 1995). In the Great Basin, island biogeography may help explain the composition of montane faunas and predict how those faunas will respond to climate change (MacArthur and Wilson 1967; Diamond 1975; Shafer 1990; McDonald and Brown 1992; Murphy and Weiss 1992; Boggs and Murphy 1997). Because mean air temperature decreases 0.6°C with every 100-meter increase in elevation, a 3°C rise in average temperature might require a species to shift its distribution upward 500 meters in order to track a specific thermal environment (Schneider et al. 1992). Forecasts about the effects of climate change in the Great Basin usually have assumed that (1) regional temperature will warm by roughly 3°C, (2) vegetation zones will shift upward by 500 meters, thereby decreasing in area, and (3) animals that are closely associated with particular vegetation zones likewise will move upward by 500 meters (McDonald and Brown 1992; Murphy and Weiss 1992; Fleishman et al. 1998). Murphy and Weiss (1992), for example, estimated the number of butterfly species that would be extirpated by climate change on the basis of the species' associations with vegetation zones (e.g., pinyon-juniper, alpine). Similarly, McDonald and Brown (1992) predicted that a 3°C rise in temperature would lead to the extirpation of 9–62 percent of the montane mammals in various Great Basin mountain ranges and 21 percent of the mammals in the Great Basin as a whole.

Clearly, the vegetation-based scenario above simplifies both climatic changes and biological responses to those changes. To some extent, oversimplification of climate change scenarios is necessary to accommodate scientific uncertainty and model tractability (Sala et al. 2000). Nonetheless, several caveats are worth noting. First, although resources for some animals are associated with a particular thermal zone, others have a comparatively opportunistic distribution. The occurrence of a particular plant species, for instance, may be driven more by availability of water or certain soils than by elevation or temperature per se. Second, plant species have individual responses to climate change: a vegetational community does not move en masse (Gleason 1926; Huntley 1991; Tausch et al. 1993; Guisan et al. 1995; Risser 1995; Kupfer and Cairns 1996). As a result, it may not be possible to predict how resource distributions for various animals will change on the basis of predicted shifts in vegetation zones. Third, animals often require resources in addition to plants.

Relatively recent data from several taxonomic groups suggest that it may be appropriate to revise or expand earlier paradigms of Great Basin biogeography. For example, using an expanded set of data on mammals in the Great Basin, Lawlor (1998) found that present-day assemblages are considerably more dynamic than previously understood. Similarly, Fleishman et al. (2001a) found that while most species of butterflies in the Great Basin may persist at the regional level, the number of butterfly extirpations at the mountain range level may vary considerably among ranges. In ranges with an average crest elevation below about 3,000 meters, the magnitude of losses may depend on whether butterflies can exploit isolated, high-elevation peaks.

During the Middle Holocene, approximately 8,000–5,000YBP, temperatures in the Great Basin were several degrees warmer than today (Van Devender et al. 1987). Thus, it might be expected that most of the montane species that currently inhabit the Great Basin would be able to tolerate the magnitude of climatic warming forecast over the next several centuries; species that were extremely sensitive to the effects of increased temperatures may already have been extirpated. However, it is not clear whether the Middle Holocene warming, which was caused by changes in solar insolation and accompanied by increases in summer precipitation, will be comparable to projected patterns of climate change (Grayson 2000). In addition, faunal responses to climate change may depend in part upon the speed at which those changes occur and the extent to which not only the mean but also the variance in climate parameters increases (McLaughlin et al. 2002).

Management and Restoration Implications

The Great Basin is not a simple or easy-to-generalize system. Instead, as demonstrated throughout this volume, ongoing studies across diverse scales are revealing numerous complexities and exceptions that defy easy classification or management. Work on the fauna of the Great Basin strongly indicates that it is possible to increase the effectiveness of maintenance and restoration efforts—whether aimed at native species or ecosystem condition more broadly—by evaluating both species-level attributes and ecosystem-scale system attributes. In many cases, as demonstrated in this chapter, exploring species responses to physical environmental variables is a powerful, practical method for explaining why faunal distributions may have shifted in the past—and for predicting how those distributions may be affected by future ecological changes, whether natural or anthropogenic. Meeting the wide range of management goals in the Great Basin, as in any ecosystem, requires adoption of an interdisciplinary approach that considers species diversity patterns in concert with ecological processes.

Synthesis of historical and contemporary data on faunal distributions, vegetation, and hydrology may contribute meaningfully to development of holistic management and conservation strategies. For example, there is considerable debate about the extent to which species diversity per se affects ecosystem processes such as primary productivity and nutrient cycling (Pimm 1991; Tilman 1999; Waide et al. 1999; Loreau 2000; Mittelbach et al. 2001). Some authors have proposed that many species are functionally redundant—that they serve the same purpose with respect to ecosystem processes (Walker 1992). If so, species richness may provide evolutionary "insurance" in the event of long-term environmental change (Yachi and Loreau 1999; Tilman 2000; Loreau et al. 2001). In the short term, however, the number of species may be less critical than the functional ecological role that each species performs (Tilman et al. 2001).

Investigation of the functional role of species may prove particularly relevant to comprehensive restoration efforts. In many situations, revegetation and management of water flows will largely be sufficient to restore major ecological processes. Occasionally, however, a particular species affects key ecological processes to an extent that greatly exceeds what would be predicted from its abundance or biomass (Mills et al. 1993; Fauth 1999). Maintenance of such "keystone" species (see table 8.1) may be critical to achieving restoration success.

Because large-scale disturbance regimes also help to maintain ecological processes, reinstating those perturbations often is considered to be a major restoration goal in its own right. Maintaining viable populations of some native species also may require reestablishment of natural cycles of fire (Covington et al. 1997; Fulé and Covington 1999) and flooding (Lake 1995; Smith 1998; Meretsky et al. 2000; Richter and Richter 2000). Clearly, manipulation of disturbance cycles, let alone restoration of historical patterns, can be prohibitively expensive and politically contentious. Although actions taken to return disturbance processes to an ecosystem may have variable success with respect to the status of target species (Sher et al. 2000; Walters et al. 2000; Swengel 2001; Gabbe et al. 2002), those actions have considerable ecological benefits at other levels. Efforts to understand interactions between the composition, structure, and function of ecosystems not only inform ecological theory, but are central to the development of effective management frameworks and strategies for ecosystem maintenance and restoration.

Acknowledgments

Many thanks to Hugh Britten, Jeanne Chambers, Ralph Mac Nally, and A. Thomas Vawter for comments that improved this chapter. Thanks also to Karen Klitz and Peter Goin for gracious assistance with photographs. Partial support for this work was provided by the Joint Fire Sciences Program via the Rocky Mountain Research Station, Forest Service, U.S. Department of Agriculture, and by the Nevada Biodiversity Research and Conservation Initiative.

LITERATURE CITED

Andelman, S. J., and W. F. Fagan. 2000. Umbrellas and flagships: Efficient conservation surrogates, or expensive mistakes? *Proceedings of the National Academy of Sciences* 97:5954–5959.

Angermeier, P. L., and J. R. Karr.1994. Biological integrity versus biological diversity as policy directives. *BioScience* 44:690–697.

Angermeier, P. L., and M. R. Winston. 1999. Characterizing fish community diversity across Virginia landscapes: Prerequisite for conservation. *Ecological Applications* 9:335–349.

Austin, M. P., A. O. Nicholls, and C. R. Margules. 1990. Measurement of the realized qualitative niche: Environmental niches of five *Eucalyptus* species. *Ecological Monographs* 60:161–177.

Bennett, A. F. 1999. *Linkages in the landscape: The roles of corridors and connectivity in wildlife conservation*. Gland, Switzerland: International Union for the Conservation of Nature.

Boggs, C. L., and D. D. Murphy. 1997. Community composition in mountain ecosystems: Climatic determinants of montane butterfly distributions. *Global Ecology and Biogeography Letters* 6:39–48.

Braithwaite, L. W., M. P. Austin, M. Clayton, J. Turner, and A. O. Nichols. 1989. On predicting the presence of birds in *Eucalyptus* forest types. *Biological Conservation* 50:33–50.

Brown, E. R., tech. ed. 1985. *Management of fish and wildlife habitats in forests of western Oregon and Washington*. USDA Forest Service General Technical Report R6-F&WL-192-1985. Portland, Ore.

Brown, J. H. 1971. Mammals on mountaintops: Non-equilibrium insular biogeography. *American Naturalist* 105:467–478.

———. 1978. The theory of insular biogeography and the distribution of boreal mammals and birds. *Great Basin Naturalist Memoirs* 2:209–228.

Busby, J. R. 1991. BIOCLIM—a bioclimatic analysis and predictive system. Pp. 64–68 in *Nature conservation: Cost-effective biological surveys and data analysis*, edited by C. R. Margules and M. P. Austin. Canberra, Australia: CSIRO.

Caicco, S. L., J. M. Scott, B. Butterfield, and B. Csuti. 1995. A gap analysis of the management status of the vegetation of Idaho (USA). *Conservation Biology* 9:498–511.

Cameron, A. C., and P. K. Trivedi. 1998. *Regression analysis of count data*. Cambridge, UK: Cambridge University Press.

Caro, T. M., and G. O'Doherty. 1999. On the use of surrogate species in conservation biology. *Conservation Biology* 13:805–814.

Carroll, S. S., and D. L. Pearson. 1998. Spatial modeling of butterfly species richness using tiger beetles (Cicindelidae) as a bioindicator taxon. *Ecological Applications* 8:531–543.

Chevan, A., and M. Sutherland. 1991. Hierarchical partitioning. *American Statistician* 45:90–96.

Christopher, R. D., and M. Fugate. 2001. *Branchinecta hiberna*, a new species of fairy shrimp (Crustacea: Anostraca) from western North America. *Western North American Naturalist* 61:11–18.

Cody, M. L. 1986. Diversity, rarity, and conservation in Mediterranean-climate regions. Pages 122–152 in *Conservation biology: The science of scarcity and diversity*, edited by M. E. Soulé. Sunderland, Mass.: Sinauer.

Coleman, F. C., and S. L. Williams. 2002. Overexploiting marine ecosystem engineers: Potential consequences for biodiversity. *Trends in Ecology and Evolution* 17:40–44.

Cooper, S. D., S. Diehl, K. Kratz, and O. Sarnelle. 1998. Implications of scale for patterns and processes in stream ecology. *Australian Journal of Ecology* 23:27–40.

Covington, W. W., P. Z. Fulé, M. M. Moore, S. C. Hart, T. E. Kolb, J. N. Mast, S. S. Sackett, and M. R. Wagner. 1997. Restoring ecosystem health in ponderosa pine forests of the southwest. *Journal of Forestry* 95:23–29.

Dennis, R. L. H. 1993. *Butterflies and climate change*. Manchester, UK: Manchester University Press.

Dennis, R. L. H., and T. G. Shreeve. 1989. Butterfly wing morphology variation in the British Isles: The influence of climate, behavioural posture and the hostplant-habitat. *Biological Journal of the Linnaean Society* 38:323–348.

Diamond, J. 1975. The island dilemma: Lessons of modern biogeographic studies for the design of nature reserves. *Biological Conservation* 7:129–146.

Dobkin, D. S., and B. A. Wilcox. 1986. Analysis of natural forest fragments: Riparian birds in the Toiyabe Mountains, Nevada. Pp. 293–299 in *Wildlife 2000: Modeling habitat relationships of terrestrial vertebrates*, edited by J. Verner, M. L. Morrison, and C. J. Ralph. Madison: University of Wisconsin Press.

Dunham, J. B., S. B. Adams, R. E. Schroeter, and D. C. Novinger. 2002a. Alien invasions in aquatic ecosystems: Toward an understanding of brook trout invasions and potential impacts on inland cutthroat trout in western North America. *Reviews in Fish Biology and Fisheries* 12:373–391.

Dunham, J. B., M. M. Peacock, B. E. Rieman, R. E. Schroeter, and G. L. Vinyard. 1999. Local and geographic variability in the distribution of stream-living Lahontan cutthroat trout. *Transactions of the American Fisheries Society* 128:875–889.

Dunham, J. B., B. E. Rieman, and J. T. Peterson. 2002b. Patch-based models of species occurrence: Lessons from salmonid fishes in streams. Pp. 327–334 in *Predicting species occurrences: Issues of scale and accuracy*, edited by J. M. Scott, P. J. Heglund, M. L. Morrison, M. G. Raphael, J. B. Haufler, and F. B. Wall. Washington, D.C.: Island Press.

Dunham, J. B., R. E. Schroeter, and B. E. Rieman. 2003. Influence of maximum water temperature on occurrence of Lahontan cutthroat trout within streams. *North American Journal of Fisheries Management* 23:1042–1049.

Dunham, J. B., G. L. Vinyard, and B. E. Rieman. 1997. Habitat fragmentation and extinction risk of Lahontan cutthroat trout (*Oncorhynchus clarki henshawi*). *North American Journal of Fisheries Management* 17:910–917.

Ehrlich, P. R., D. S. Dobkin, and D. Whey. 1988. *The birder's handbook*. New York: Simon and Schuster.

Fagan, W. F., and P. M. Kareiva. 1997. Using compiled species lists to make biodiversity comparisons among regions: A test case using Oregon butterflies. *Biological Conservation* 80:249–259.

Fauth, J. E. 1999. Identifying potential keystone species from field data—an example from temporary ponds. *Ecology Letters* 2:36–43.

Fleishman, E., G. T. Austin, and D. D. Murphy. 1997. Natural history and biogeography of the butterflies of the Toiyabe Range, Nevada (Lepidoptera: Papilionoidea). *Holarctic Lepidoptera* 4:1–18.

———. 2001a. Biogeography of Great Basin butterflies: Revisiting patterns, paradigms, and climate change scenarios. *Biological Journal of the Linnean Society* 74:501–515.

Fleishman, E., G. T. Austin, and A. D. Weiss. 1998. An empirical test of Rapoport's rule: Elevational gradients in montane butterfly communities. *Ecology* 79:2482–2493.

Fleishman, E., R. B. Blair, and D. D. Murphy. 2001c. Empirical validation of a method for umbrella species selection. *Ecological Applications* 11:1489–1501.

Fleishman, E., and R. Mac Nally. Forthcoming. Linking models of species occurrence and landscape reconstruction. *Transactions of the Western Section of the Wildlife Society.*

Fleishman, E., R. Mac Nally, J. P. Fay, and D. D. Murphy. 2001b. Modeling and predicting species occurrence using broad-scale environmental variables: An example with butterflies of the Great Basin. *Conservation Biology* 15:1674–1685.

Fleishman, E., D. D. Murphy, and P. F. Brussard. 2000. A new method for selection of umbrella species for conservation planning. *Ecological Applications* 10:569–579.

Franklin, J. F. 1988. Structural and functional diversity in temperate forests. Pp. 166–175 in *Biodiversity*, edited by E. O. Wilson. Washington, D.C.: National Academy Press.

Freitag, S., A. S. van Jaarsveld, and H. C. Biggs. 1997. Ranking priority biodiversity areas: An iterative conservation value-based approach. *Biological Conservation* 82:263–272.

Fulé, P. Z., and W. W. Covington. 1999. Fire regime changes in La Michilía Biosphere Reserve, Durango, Mexico. *Conservation Biology* 13:640–652.

Gabbe, A. P., S. K. Robinson, and J. D. Brawn. 2002. Tree-species preferences of foraging insectivorous birds: Implications for floodplain restoration. *Conservation Biology* 16:462–470.

Garth, J. S., and J. W. Tilden. 1986. *California butterflies*. Berkeley: University of California Press.

Gleason, H. A. 1926. The individualistic concept of the plant association. *Bulletin of the Torrey Botanical Club* 53:7–26.

Grayson, D. K. 2000. Mammalian responses to Middle Holocene climatic change in the Great Basin of the western United States. *Journal of Biogeography* 27:181–192.

Grayson, D. K., and S. D. Livingston. 1993. Missing mammals on Great Basin mountains: Holocene extinctions and inadequate knowledge. *Conservation Biology* 7:527–532.

Grayson, D. K., and D. B. Madson. 2000. Biogeographic implications of recent low-elevation recolonization by *Neotoma cinerea* in the Great Basin. *Journal of Mammalogy* 81:1100–1105.

Guisan, A., J. I. Holten, R. Spichiger, and L. Tessier, eds. 1995. *Potential ecological impacts of climate change in the Alps and Fennoscandian mountains*. Geneva, Switzerland: Conservatory and Botanical Garden of Geneva.

Guisan, A., and N. E. Zimmermann. 2000. Predictive habitat distribution models in ecology. *Ecological Modelling* 135:147–186.

Gustafsson, L. 2000. Red-listed species and indicators: Vascular plants in woodland key habitats and surrounding production forests in Sweden. *Biological Conservation* 92:35–43.

Hall, E. R. 1946. *Mammals of Nevada.* Berkeley: University of California Press.

Hansen, A. J., and D. L. Urban. 1992. Avian response to landscape pattern: The role of species' life histories. *Landscape Ecology* 7:163–180.

Hanski, I. 1999. *Metapopulation ecology.* New York: Oxford University Press.

Hanski, I., and M. E. Gilpin, eds. 1997. *Metapopulation biology.* San Diego, Calif.: Academic Press.

Hanski, I., A. Moilanen, T. Pakkala, and M. Kuussaari. 1996. The quantitative incidence function model and persistence of an endangered butterfly metapopulation. *Conservation Biology* 10:578–590.

Hepworth, D. K., M. J. Ottenbacher, and L. N. Berg. 1997. Distribution and abundance of native Bonneville cutthroat trout (*Oncorhynchus clarki utah*) in southwestern Utah. *Great Basin Naturalist* 57:11–20.

Hershler, R. 1998. A systematic review of the hydrobiid snails (Gastropoda: Rissooidea) of the Great Basin, western United States. Part 1. Genus *Pyrgulopsis. Veliger* 41:1–132.

———. 1999. A systematic review of the hydrobiid snails (Gastropoda: Rissooidea) of the Great Basin, western United States. Part 2. Genera *Colligyrus, Eremopyrgus, Fluminicola, Pristinicola,* and *Tryonia. Veliger* 42:306–337.

Hubbs, C. L., and R. R. Miller. 1948. The zoological evidence: Correlation between fish distribution and hydrographic history in the desert basins of western United States. Pp. 17–166 in *The Great Basin with emphasis on glacial and postglacial times.* Bulletin of University of Utah 30. Provo: University of Utah.

Huntley, B. 1991. How plants respond to climate change: Migration rates, individuals and the consequences for plant communities. *Annals of Botany* 67:15–22.

Huxel, G. R., and A. Hastings. 1999. Habitat loss, fragmentation, and restoration. *Restoration Ecology* 7:309–315.

Iftner, D. C., J. A. Shuey, and J. V. Calhoun. 1992. *Butterflies and skippers of Ohio.* Ohio Biological Survey Bulletin New Series 9(1).

Jackson, L. J., A. S. Trebitz, and K. L. Cottingham. 2000. An introduction to the practice of ecological modelling. *BioScience* 50:694–706.

Johnson, N. K. 1978. Patterns of avian geography and speciation in the Intermountain Region. *Great Basin Naturalist Memoirs* 2:137–159.

Jones, C. G. 1997. Positive and negative effects of organisms as physical ecosystem engineers. *Ecology* 78:1946–1957.

Jones, C. G., J. H. Lawton, and M. Shacak. 1994. Organisms as ecosystem engineers. *Oikos* 69:373–386.

Kingsolver, J. G. 1983. Thermoregulation and flight in *Colias* butterflies: Elevational patterns and mechanistic limitations. *Ecology* 64:534–545.

———. 1989. Weather and the population dynamics of insects: Integrating physiological and population ecology. *Physiological Zoology* 62:314–334.

Kremen, C. 1994. Biological inventory using target taxa: A case study of the butterflies of Madagascar. *Ecological Applications* 4:407–422.

Kulkoyluoglu, O. 2000. Distribution and ecology of freshwater Ostracoda (Crustacea) collected from springs of Nevada, Idaho, and Oregon: A preliminary study. *Western North American Naturalist* 60:291–303.

Kupfer, J. A., and D. M. Cairns. 1996. The suitability of montane ecotones as indicators of global climatic change. *Progress in Physical Geography* 20:253–272.

Lake, P. S. 1995. Of floods and droughts: River and stream ecosystems of Australia. Pp. 659–690 in *River and stream ecosystems* 22 edited by C. E. Cushing, K. W. Cummins, and G. W. Minshall. Amsterdam, Netherlands: Elsevier.

Lambeck, R. J. 1997. Focal species: A multi-species umbrella for nature conservation. *Conservation Biology* 11:849–856.

Landres, P. B. 1992. Ecological indicators: Panacea or liability? Pp. 1295–1318 in *Eco- ·logical indicators*, edited by D. H. McKenzie, D. E. Hyatt, and V. J. McDonald. Vol. 2. New York: Elsevier Applied Science.

Landres, P. B., J. Verner, and J. W. Thomas. 1988. Critique of vertebrate indicator species. *Conservation Biology* 2:316–328.

Lawlor, T. E. 1998. Biogeography of Great Basin mammals: Paradigm lost? *Journal of Mammalogy* 79:1111–1130.

Leader-Williams, N., and H. T. Dublin. 2000. Charismatic megafauna as "flagship species." Pp. 53–80 in *Priorities for the conservation of mammalian diversity: Has the panda had its day?* edited by A. Entwistle and N. Dunstone. Cambridge, UK: Cambridge University Press.

Lindenmayer, D. B., R. B. Cunningham, M. T. Tanton, A. P. Smith, and H. A. Nix. 1990. Habitat requirements of the mountain brushtail possum and the greater glider in the montane ash-type eucalypt forests of the central highlands of Victoria (Australia). *Australian Wildlife Research* 17:467–478.

Lindenmayer, D. B., C. R. Margules, and D. B. Botkin. 2000. Indicators of biodiversity for ecological sustainable management. *Conservation Biology* 14:941–950.

Loder, N., K. J. Gaston, and H. R. Arnold. 1998. Body size and feeding specificity: Macrolepidoptera in Britain. *Biological Journal of the Linnean Society* 63:121–139.

Lomolino, M. V. 1996. Investigating causality of nestedness of insular communities: Selective immigrations or extinctions? *Journal of Biogeography* 23:699–703.

Longino, J. T., and R. K. Colwell. 1997. Biodiversity assessment using structured in- ventory: Capturing the ant fauna of a tropical rain forest. *Ecological Applications* 7:1263–1277.

Loreau, M. 2000. Biodiversity and ecosystem functioning: Recent theoretical advances. *Oikos* 91:3–17.

Loreau, M., S. Naeem, P. Inchausti, J. Bengtsson, J. P. Grime, A. Hector, D. U. Hooper, M. A. Huston, D. Raffaelli, B. Schmid, D. Tilman, and D. A. Wardle. 2001. Biodiversity and ecosystem functioning: Current knowledge and future challenges. *Science* 294:804–808.

Mac Nally, R. 2000. Regression and model-building in conservation biology, biogeog- raphy and ecology: The distinction between—and reconciliation of—"predictive" and "explanatory" models. *Biodiversity and Conservation* 9:655–671.

Mac Nally, R., and E. Fleishman. 2002. Using "indicator" species to model species richness: Model development and predictions. *Ecological Applications* 12:79–92.

Mac Nally, R., E. Fleishman, J. P. Fay, and D. D. Murphy. 2003. Modeling butterfly species richness using mesoscale environmental variables: Model construction and validation. *Biological Conservation* 110:21–31.

MacArthur, R. H., and E. O. Wilson. 1967. *The theory of island biogeography.* Prince- ton, N.J.: Princeton University Press.

Margules, C. R., and R. L. Pressey. 2000. Systematic conservation planning. *Nature* 405:243–253.

McDonald, K. A., and J. H. Brown. 1992. Using montane mammals to model extinc- tions due to climate change. *Conservation Biology* 6:409–415.

McLaughlin, J. F., J. J. Hellmann, C. L. Boggs, and P. R. Ehrlich. 2002. Climate change hastens population extinctions. *Proceedings of the National Academy of Sciences* (USA) 99:6070–6074.

Meffe, G. K., and C. R. Carroll. 1994. *Principles of conservation biology.* Sunderland, Mass.: Sinauer.

Meretsky, V. J., D. L. Wegner, and L. E. Stevens. 2000. Balancing endangered species and ecosystems: A case study of adaptive management in Grand Canyon. *Environmental Management* 25:579–586.

Miller, J. R., and P. Cale. 2000. Behavioral mechanisms and habitat use by birds in a fragmented agricultural landscape. *Ecological Applications* 10:1732–1748.

Mills, L. S., M. E. Soulé, and D. F. Doak. 1993. The keystone-species concept in ecology and conservation. *BioScience* 43:219–224.

Minckley, W. L., and J. E. Deacon. 1991. Western fishes and the real world: The enigma of endangered species revisited. Pp. 405–414 in *Battle against extinction: Native fish management in the American West,* edited by W. L. Minckley and J. E. Deacon. Tucson: University of Arizona Press.

Mittelbach, G. G., C. F. Steiner, S. M. Scheiner, K. L. Gross, H. L. Reynolds, R. B. Waide, M. R. Willig, S. I. Dodson, and L. Gough. 2001. What is the observed relationship between species richness and productivity? *Ecology* 82:2381–2396.

Moilanen, A. 2000. The equilibrium assumption in estimating the parameters of metapopulation models. *Journal of Animal Ecology* 69:143–153.

Murphy, D. D., and S. B. Weiss. 1992. Effects of climate change on biological diversity in western North America: Species losses and mechanisms. Pp. 355–268 in *Global warming and biological diversity,* edited by R. L. Peters and T. E. Lovejoy. New Haven, Conn.: Yale University Press.

Myers, N. 1986. Tropical deforestation and a mega-extinction spasm. Pp. 394–409 in *Conservation biology: The science of scarcity and diversity,* edited by M. E. Soulé. Sunderland, Mass.: Sinauer.

Neave, H. M., and T. W. Norton. 1998. Biological inventory for conservation evaluation 4: Composition, distribution and spatial prediction of vegetation assemblages in southern Australia. *Forest Ecology and Management* 106:259–281.

Niemi, G. J., J. M. Hanowski, A. R. Lima, T. Nicholls, and N. Weiland. 1997. A critical analysis on the use of indicator species in management. *Journal of Wildlife Management* 61:1240–1252.

Noss, R. F. 1990. Indicators for monitoring biodiversity: A hierarchial approach. *Conservation Biology* 4:355–364.

Oliver, I., and A. J. Beattie. 1996. Designing a cost-effective invertebrate survey: A test of methods for rapid assessment of biodiversity. *Ecological Applications* 6:594–607.

Palik, B. J., P. C. Goebel, L. K. Kirkman, and L. West. 2000. Using landscape hierarchies to guide restoration of disturbed ecosystems. *Ecological Applications* 10:189–202.

Pearson, D. L. 1994. Selecting indicator taxa for the quantitative assessment of biodiversity. *Philosophical Transactions of the Royal Society of London. Series B, Biological Sciences* 345:75–79.

Pimm, S. L. 1991. *The balance of nature?* Chicago: University of Chicago Press.

Pollard, E. 1988. Temperature, rainfall and butterfly numbers. *Journal of Applied Ecology* 25:819–828.

Prendergast, J. R. 1997. Species richness covariance in higher taxa—empirical tests of the biodiversity indicator concept. *Ecography* 20:210–216.

Reichman, O. J., and E. W. Seabloom. 2002. The role of pocket gophers as subterranean ecosystem engineers. *Trends in Ecology and Evolution* 17:44–49.

Richter, B. D., and H. E. Richter. 2000. Prescribing flood regimes to sustain riparian ecosystems along meandering rivers. *Conservation Biology* 14:1467–1478.

Ricketts, T. H., E. Dinerstein, D. M. Olson, and C. Loucks. 1999. Who's where in North America? *BioScience* 49:369–381.

Ricklefs, R., and D. Schluter. 1993. *Species diversity in ecological communities.* Chicago: University of Chicago Press.

Rieman, B. E., P. F. Hessburg, D. C. Lee, R. F. Thurow, and J. R. Sedell. 2000. Toward an integrated classification of ecosystems: Defining opportunities for managing fish and forest health. *Environmental Management* 25:425–444.

Risser, P. G. 1995. The status of the science examining ecotones. *BioScience* 45:318–325.

Rosenzweig, M. L. 1995. *Species diversity in space and time.* Cambridge, UK: Cambridge University Press.

Rubinoff, D. 2001. Evaluating the California gnatcatcher as an umbrella species for conservation of southern California coastal sage scrub. *Conservation Biology* 15:1374–1383.

Sada, D. W., J. E. Williams, J. C. Silvey, A. Halford, J. Ramakka, P. Summers, and L. Lewis. 2001. *Riparian area management: A guide to managing, restoring, and conserving springs in the western United States.* Technical Reference 1737-17. Bureau of Land Management. Denver, Colo.

Sala, O. E., F. S. Chapin, J. J. Armesto, E. Berlow, J. Bloomfield, R. Dirzo, E. Huber-Sanwald, L. F. Huenneke, R. B. Jackson, A. Kinzig, R. Leemans, D. M. Lodge, H. A. Mooney, M. Oesterheld, N. L. Poff, M. T. Sykes, B. H. Walker, M. Walker, and D. H. Wall. 2000. Global biodiversity scenarios for the year 2100. *Science* 287:1770–1774.

Schneider, S. H., L. Mearns, and P. H. Gleick. 1992. Climate-change scenarios for impact assessment. Pp. 38–55 in *Global warming and biological diversity,* edited by R. L. Peters and T. E. Lovejoy. New Haven, Conn.: Yale University Press.

Schwarz, G. 1978. Estimating the dimension of a model. *Annals of Statistics* 6:461–464.

Scott, C. T. 1998. Sampling methods for estimating change in forest resources. *Ecological Applications* 8:228–233.

Scott, J. A. 1986. *The butterflies of North America.* Stanford, Calif.: Stanford University Press.

Scott, J. M., P. J. Heglund, M. L. Morrison, M. G. Raphael, J. B. Haufler, and F. B. Wall, eds. 2002. *Predicting species occurrences: Issues of scale and accuracy.* Washington, D.C.: Island Press.

Shafer, C. L. 1990. *Nature reserves, island theory and conservation practice.* Washington, D.C.: Smithsonian Institution Press.

Sher, A. A., D. L. Marshall, and S. A. Gilbert. 2000. Competition between native *Populus deltoides* and invasive *Tamarix ramosissima* and the implications for reestablishing flooding disturbance. *Conservation Biology* 14:1744–1754.

Smith, D. I. 1998. *Water in Australia: Resources and management.* Oxford, UK: Oxford University Press.

Smith, G. R. 1978. Biogeography of intermountain fishes. *Great Basin Naturalist Memoirs* 2:17–42.

Smith, S. D., T. E. Huxman, S. F. Zitzer, T. N. Charlet, D. C. Housman, J. S. Coleman, L. K. Fenstermaker, J. R. Seemann, and R. S. Nowak. 2000. Elevated CO_2 increases productivity and invasive species success in an arid ecosystem. *Nature* 408:79–82.

Springer, P., and C. L. Boggs. 1986. Resource allocation to oocytes: Heritable variation with altitude in *Colias philodice eriphyle* (Lepidoptera). *American Naturalist* 127:252–256.

Stohlgren, T. J., J. F. Quinn, M. Ruggiero, and G. S. Waggoner. 1995. Status of biotic inventories in U.S. national parks. *Biological Conservation* 71:97–106.

Swengel, A. B. 2001. A literature review of insect responses to fire, compared to other conservation managements of open habitat. *Biodiversity and Conservation* 10:1141–1169.

Tausch, R. J., P. E. Wigand, and B. Burkhardt. 1993. Viewpoint: Plant community thresholds, multiple steady states, and multiple successional pathways: Legacy of the Quaternary. *Journal of Range Management* 46:439–447.

Thomas, J. A. 1995. The conservation of declining butterfly populations in Britain and Europe: Priorities, problems and successes. *Biological Journal of the Linnean Society* 56(Supplement):55–72.

Tilman, D. 1999. The ecological consequences of changes in biodiversity: A search for general principles. *Ecology* 80:1455–1474.

———. 2000. Causes, consequences and ethics of biodiversity. *Nature* 405:208–211.

Tilman, D., P. B. Reich, J. Knops, D. Wedin, T. Mielke, and C. Lehman. 2001. Diversity and productivity in a long-term grassland experiment. *Science* 294:843–845.

U.S. EPA (United States Environmental Protection Agency). 1999. *Climate change and Nevada.* Publication EPA 236-F-98-007. www.epa.gov/globalwarming/impacts/stateimp/nevada/index.html.

Van Devender, T. R., R. S. Thompson, and J. L. Betancourt. 1987. Vegetation history of the deserts of southwestern North America: The nature and timing of the Late Wisconsin—Holocene transition. Pp. 323–352 in *North America and adjacent oceans during the last deglaciation*, edited by W. F. Ruddiman and H. E. Wright Jr. Geological Society of North America. Boulder, Colo.

Waide, R. B., M. R. Willig, C. F. Steiner, G. Mittelbach, L. Gough, S. I. Dodson, G. P. Juday, and R. Parmenter. 1999. The relationship between productivity and species richness. *Annual Review of Ecology and Systematics* 30:257–300.

Walker, B. 1992. Biodiversity and ecological redundancy. *Conservation Biology* 6:18–23.

Walters, C., J. Korman, L. E. Stevens, and B. Gold. 2000. Ecosystem modeling for evaluation of adaptive management policies in the Grand Canyon. *Conservation Ecology* 4(2). http://www.consecol.org/Journal/.

Weaver, J. S., and M. J. Myers. 1998. Two new species of caddisflies of the genus *Lepidostoma* Rambur (Trichoptera: Lepidostomatidae) from the Great Basin. *Aquatic Insects* 20:189–195.

Wiens, J. A. 1989. Spatial scaling in ecology. *Functional Ecology* 3:385–397.

Wilcox, B. A. 1980. Insular ecology and conservation. Pp. 95–117 in *Conservation Biology: An evolutionary-ecological perspective*, edited by M. E. Soulé and B. A. Wilcox. Sunderland, Mass.: Sinauer.

Wilcox, B. A., D. D. Murphy, P. R. Ehrlich, and G. T. Austin. 1986. Insular biogeography of the montane butterfly fauna in the Great Basin: Comparison with birds and mammals. *Oecologia* 69:188–194.

Yachi, S., and M. Loreau. 1991. Biodiversity and ecosystem productivity in a fluctuating environment: The insurance hypothesis. *Proceedings of the National Academy of Sciences* 96:1463–1468.

Young, M. K., ed. 1995. *Conservation assessment for inland cutthroat trout.* General Technical Report RM-256, UDSA Forest Service, Rocky Mountain Research Station. Fort Collins, Colo.

Zacharias, M. A., and J. C. Roff. 2001. Use of focal species in marine conservation and management: A review and critique. Aquatic Conservation: *Marine and Freshwater Ecosystems* 11:59–76.

Chapter 9

Process-Based Approaches for Managing and Restoring Riparian Ecosystems

JEANNE C. CHAMBERS, JERRY R. MILLER,
DRU GERMANOSKI, AND DAVE A. WEIXELMAN

The success of riparian restoration rests on understanding the underlying processes structuring the ecosystems of interest at the appropriate temporal and spatial scales and on understanding the effects of historical and ongoing land uses (Williams et al. 1997; National Research Council 1992, 2002). The Great Basin Ecosystem Management Project (EM Project) has used a two-pronged approach to establish the necessary foundation for restoring and maintaining riparian areas in upland watersheds of the central Great Basin. First, it provides an understanding of the underlying abiotic and biotic processes structuring the riparian areas. Second, it evaluates the long- and short-term effects of climate change and natural and anthropogenic disturbance on these processes. As described in previous chapters, the EM Project has examined the geomorphic, hydrologic, and biotic processes influencing riparian areas in the central Great Basin over temporal scales ranging from the mid-Holocene to the present, and spatial scales ranging from entire watersheds to localized stream reaches. The EM Project's integrated analyses of the geomorphic, hydrologic, and biotic processes have supplied the requisite information for developing a process-based and multiscale approach for maintaining or restoring sustainable riparian ecosystems.

In this chapter, an overview is provided of our current understanding of the abiotic and biotic processes structuring riparian ecosystems in the central Great Basin. A conceptual basis for managing and restoring sustainable streams and riparian ecosystems is then presented based on our knowledge of both current processes and ecosystem trajectories. The remainder of the chapter focuses on actual management and restoration strategies for these ecosystems. This approach addresses the major physical and biotic problems inherent within central Great Basin watersheds and is meant to serve as a model for linking process-based studies with man-

261

agement and restoration activities. It is a multiscale approach that includes the watershed/riparian corridor and valley segment/stream reach.

Processes Structuring Great Basin Riparian Areas

The processes currently structuring riparian areas in the central Great Basin are strongly influenced by past climates. The paleoecological records collected as part of the EM Project, as well as previous investigations, indicate that a major drought occurred in the region from approximately 2500 to 1300 YBP (Miller et al. 2001; chapter 2). During this drought, most of the available sediments were stripped from the hillslopes and deposited on the valley floors and on side-valley alluvial fans (chapter 3). As a consequence of this hillslope erosion, streams are now sediment limited and have a natural tendency to incise. In fact, the geomorphic data indicate that over the past two thousand years, the dominant response of the streams to disturbance has been incision. The most recent episode of incision began about 500 to 400 YBP, which predates Anglo-American settlement of the area in the 1860s.

The tendency of a stream to incise depends on the sensitivity of the watershed to both natural and anthropogenic disturbance. Analyses of central Great Basin watersheds indicate that stream incision is closely related to watershed characteristics, including geology, size, and morphology, and to valley segment attributes like gradient, width, and substrate size (chapter 4). In these semiarid ecosystems where precipitation and, thus, streamflow is highly variable both between and within years, most incision occurs during episodic, high-flow events. Since the initiation of the EM Project in 1994, high-flow events capable of producing significant incision have occurred in 1995 and 1998 (Chambers et al. 1998; Germanoski et al. 2001). Watersheds that are highly sensitive to disturbance respond to more-frequent, lower-magnitude runoff events than watersheds that are less sensitive to disturbance. The combined geomorphic and hydrologic characteristics of the watersheds determine the composition and pattern of riparian vegetation at watershed- to valley-segment scales (chapter 7). Thus, watershed attributes that characterize basin sensitivity to disturbance also have good predictive value for vegetation types and associations.

A number of the watersheds are characterized by prominent side-valley alluvial fans that influence both stream and riparian ecosystems. The fans reached their maximum extent during the drought that occurred from about 2500 to 1300 YPB (Miller et al. 2001; chapter 3). Watersheds with well-developed fans often are characterized by stepped-valley profiles and, consequently, riparian corridors that exhibit abrupt changes in local geomorphic and hydrologic attributes. Because of the relationships among geomorphic characteristics, hydrologic regimes, and riparian vegetation, the fans also influence ecosystem patterns within the riparian corridors (Korfmacher 2001; chapters 4 and 7). Many of the fans currently serve as local base-level controls that determine the rate and magnitude of upstream incision.

Riparian ecosystems located immediately upstream of alluvial fans often are at risk of stream incision through the fan deposits. Many of the fans have multiple knickpoints (a short, oversteepened segment of the longitudinal profile of the channel) and are subject to stream incision due to high shear stress associated with knickzone migration during high-flow events. Similarly, ecosystems located in watersheds with pseudostable channels can be degraded during catastrophic incision via groundwater sapping. Because the stream channel serves as a groundwater discharge point, it represents the base level to which the hydraulic gradient of the groundwater system is adjusted. Stream incision lowers this base level for groundwater discharge, resulting in declines in water table levels and subsequent changes in the composition and structure of riparian vegetation (Castelli et al. 2000; Wright and Chambers 2002). Meadow complexes are presently the ecosystems most susceptible to degradation not only because they are often located upstream of alluvial fans, but also because they are subject to processes such as groundwater sapping (chapter 5).

The rate and magnitude of steam incision in central Great Basin watersheds have been increased by anthropogenic disturbances. Because most of the streams have been prone to incision for the past two thousand years, separating changes attributable to ongoing stream incision from those associated with anthropogenic disturbance can be exceedingly complex (chapter 3). In the cases of roads, diversions, and livestock or recreational trails, the point of initiation of stream incision often can be identified and the local effects of the disturbance on the stream reconstructed. The most direct evidence that anthropogenic disturbance has influenced stream incision in the central Great Basin is derived from ongoing studies on the effects of roads on riparian areas. Increasingly, roads are identified as major causes of stream incision and riparian area degradation across the United States (USDA Forest Service 1997; Forman and Deblinger 2000; Trombulak and Frissel 2000). In the central Great Basin, several cases of "road captures" have been documented and many others observed where streams have been diverted onto road surfaces during high flows (Lahde 2003).

This diversion has resulted in increased shear stress and stream power and, ultimately, stream incision (Lahde 2003). Once initiated, stream incision often continues to occur as a result of knickpoint migration.

Assigning cause and effect to more diffuse anthropogenic disturbances such as overgrazing by livestock is more difficult. In general, overgrazing by livestock can negatively affect stream bank and channel stability, and localized changes in stream morphology often have been associated with overgrazing by livestock in the western United States (see reviews in Trimble and Mendel 1995; Belsky et al. 1999). However, data that clearly demonstrate the relationship(s) between regional stream incision and overgrazing by livestock have not been collected for the central Great Basin. In reality, it may never be possible to precisely distinguish the amount of channel incision caused by climate change from that due to anthropogenic disturbance. From a restoration and management standpoint, it is important to recognize that because particular types of streams are prone to incision, they have greater sensitivity to both natural and anthropogenic disturbances.

Anthropogenic disturbances have effects on central Great Basin riparian ecosystems that are unrelated to stream incision. In semiarid rangelands, like those in the western United States, the general degradation of riparian areas has been attributed primarily to overgrazing by livestock (see reviews in Kauffman and Krueger 1984; Clary and Webster 1989; Skovlin 1984; Fleischner 1994; Ohmart 1996; Belsky et al. 1999). Livestock grazing influences riparian ecosystems by (1) removing herbage, which allows soil temperatures to rise and results in increased evaporation; (2) damaging plants due to rubbing, trampling, grazing, or browsing; (3) altering nutrient dynamics by depositing nitrogen in excreta from animals and removing foliage; and (4) compacting soil, which increases runoff and decreases water availability to plants. Research conducted on Great Basin meadow ecosystems (Weixelman et al. 1997; chapter 7) and elsewhere shows that these effects can cause changes in plant physiology, population dynamics, and community attributes such as cover, biomass, composition, and structure. In addition, overgrazing by livestock can result in local decreases in water quality as a result of increased nitrogen and phosphorus inputs via urine and feces, and higher sediment inputs due to bank trampling and increased bank failure (Sidle and Amacher 1990). Roads also have negative effects on water quality as a result of increased sediment from road drainage and road crossings (M. Amacher, unpublished data). Disturbances due to recreation, such as campsites, and vehicle and foot traffic, have increasingly widespread effects but have been poorly documented.

The cumulative effects of climatic perturbation and anthropogenic disturbance in the central Great Basin have multiple consequences for streams and their associated riparian ecosystems. There have been major changes in channel pattern and form and many streams have been isolated from their floodplains (chapters 3 and 4). Surface water and groundwater interactions have been altered (Germanoski et al. 2001; chapter 5), and declines in water tables have caused changes in plant species composition and vegetation structure (Wright and Chambers 2002; chapter 7). Overgrazing by livestock and other anthropogenic disturbances have caused additional degradation of these ecosystems. The net effects of these changes have been a decrease in the areal extent of riparian corridors and a reduction in habitat quantity and quality for both aquatic and terrestrial animals.

Conceptual Basis for Management and Restoration

The importance of developing a conceptual basis for managing and restoring degraded ecosystems is gaining increasing recognition (Allen et al. 1997; Williams et al. 1997; Whisenant 1999; National Research Council 2002). For degraded riparian ecosystems like those in the Great Basin, such a conceptual basis must consider both the type and characteristics of degradation, and the recovery potential as determined by the underlying physical and biotic processes.

In the central Great Basin many of the streams and, consequently, riparian ecosystems are currently in a nonequilibrium state. Because of the drought that occurred between 2500 and 1300 YBP and the erosion of available hillslope sediments, the streams are sediment limited and exhibit a tendency to incise. Some of the streams have adjusted to the new set of geomorphic conditions, in other words, they have reached their maximum depth of incision under the current sediment and hydrologic regime. Others are still adjusting and will continue to incise because of heterogeneous channel profiles and the lack of hillslope sediments. Due to the resulting changes in stream processes and surface and groundwater relations, riparian ecosystems have crossed thresholds. For our purposes, threshold crossings occur when a system does not return to the original state following disturbance (Ritter et al. 1999). Threshold crossings can be defined based on the limits of natural variability within systems. For streams and riparian ecosystems that have crossed geomorphic and hydrologic thresholds, returning the system to a predisturbance state is an unrealistic goal. Thus, it is necessary to base concepts of sustainability and approaches to management on current, and not historic, stream processes and riparian ecosystem conditions.

As outlined in chapter 1, the goal of restoration and management activities in the central Great Basin is sustainable stream and riparian ecosystems. Sustainable ecosystems, over the normal cycle of disturbance events, retain characteristic processes including rates and magnitudes of geomorphic activity, hydrologic flux and storage, biogeochemical cycling and storage, and biological activity and production (modified from Chapin et al. 1996 and Christensen et al. 1996). Sustainable stream and riparian ecosystems also exhibit physical, chemical, and biological linkages among their geomorphic, hydrologic, and biotic components (Gregory et al. 1991). Thus, for the purposes of this volume, managing and restoring riparian areas is defined as reestablishing or maintaining sustainable fluvial systems and riparian ecosystems that exhibit both characteristic processes and related biological, chemical, and physical linkages among system components (modified from National Research Council 1992). Inherent in this definition is the idea that sustainable ecosystems provide important ecosystem services. In the central Great Basin, ecosystem services from riparian areas include an adequate supply of high-quality water, habitat for a diverse array of aquatic and terrestrial organisms, forage and browse for native herbivores and livestock, and recreational opportunities.

Restoration and Management Approaches

The importance of restoration approaches that address all scales but that begin at the level of the watershed is increasingly emphasized by the scientific community (Williams et al. 1997; National Research Council 2002). In the following sections, an understanding of stream and riparian ecosystem processes in the central Great Basin is used to develop restoration and management approaches for the watershed/riparian corridor and valley segment/stream reach scales. Our focus is on generalized approaches that address the major physical and biotic concerns in the watersheds rather than on methods that are specific to a single disturbance or site type.

Watershed/Riparian-Corridor Scale

At the scale of the watershed/riparian corridor, a realistic restoration and management goal is to restore and maintain stable stream and riparian ecosystems and to slow the rates of change in incising systems, especially where anthropogenic disturbance is accelerating degradation. An important first step in meeting this goal is to understand the sensitivity or response of individual watersheds to disturbance. This approach increases our ability to predict future responses to both disturbance and management, and can be used to (1) determine restoration potentials, and (2) develop appropriate management schemes. Ecologists evaluate the sensitivity of ecosystems to disturbance on the basis of their resistance and resilience. As used here, *resistance* refers to the ability of an ecosystem to maintain characteristic processes despite various stressors or disturbances, while *resilience* refers to the capacity to regain those same processes over time following stress or disturbance (Society for Ecological Restoration International 2002). Geomorphologists have described the sensitivity of landforms to disturbance as the propensity for a change in the environment to lead to a new equilibrium state (Schumm and Brakenridge 1987; chapter 4). The sensitivity of watersheds to disturbances is related to the interaction of driving and resisting forces that regulate erosional and depositional processes. Thus, it is controlled by factors such as the erosional resistance of the underlying bedrock and channel-forming materials, basin relief and morphometry (drainage density, stream frequency, Shreve magnitude), and watershed hydrology (Gerrard 1993).

As part of the EM Project, a categorization of central Great Basin watersheds has been developed that is based on their sensitivity to past and present disturbance (chapter 4). This categorization utilizes several parameters that provide insights into the magnitude, timing, and nature of channel incision along riparian corridors, including depth of channel incision; number and continuity of terraces; channel gradient and uniformity, including the influence of fans and presence of knickpoints; erosion uniformity along the channel bed; and the magnitude of recent erosional/depositional events. Analysis of the indicator variables illustrates that all of the channels included in the categorization exhibit some degree of incision but that distinct differences exist in the nature, timing, and rate of response to disturbance. Thus, it is possible to categorize the watersheds according to measurable parameters related to stream incision. However, certain assumptions are necessary for the actual categorization of watershed sensitivity: (1) the most sensitive basins are those that respond first to disturbance, (2) channels that are the most responsive to a disturbance or high-magnitude runoff event are sensitive, and (3) channels that have maintained vestiges of their predisturbance morphology, and that are continuing to adjust, are less sensitive. Using this approach, four groups of basins have been described: Group 1, flood dominated; Group 2, deeply incised; Group 3, fan dominated; and Group 4, pseudostable. It is important to recognize that these categories were developed using indicator variables that specify the nature of the responses in upland watersheds of the central Great Basin to disturbance and that extrapolating the categorization to other regions characterized by different geologic and geomorphic conditions may lead to erroneous results. Nevertheless, the *approach* that was used is applicable elsewhere and may provide the necessary information on the abiotic and biotic processes within watersheds to develop sound restoration and management initiatives.

Riparian plant species and vegetation types are closely related to the geomorphic characteristics of the watersheds. Moreover, the variables used to categorize Great Basin watersheds according to disturbance sensitivity are useful predictors of riparian vegetation. In a set of complementary studies, the effects of watershed and channel characteristics, including the presence of alluvial fans, on the composition and patterns of riparian vegetation were examined (Korfmacher 2001; chapter 7). In one of the investigations, relationships among stream channel characteristics and riparian vegetation for uniform valley segments were assessed within watersheds that had differing geology and morphometry. The other study evaluated the influence of side-valley alluvial fans on both the geomorphic characteristics and the riparian vegetation of riparian corridors. Watersheds with different geomorphic characteristics and sensitivities to disturbance are characterized by unique vegetation types as has been demonstrated for other semiarid ecosystems (Harris 1988). Vegetation types are related to individual watershed characteristics, including terrace height and number, channel gradient, channel and bank substrate size, channel width/depth ratio, and incised channel depth (chapter 7). Side-valley alluvial fans influence the geomorphic (chapter 3) and hydrologic (chapter 5) characteristics of riparian corridors and, thus, vegetation patterning (Korfmacher 2001; chapter 7). Geomorphic position relative to side-valley alluvial fans (i.e., upstream, adjacent to, or below the fan) determines valley width, valley floor lateral slope, and bankfull depth and influences the relative proportions of meadow, woody, and upland vegetation types.

Combining the categorization of watershed sensitivity (chapter 4) with our understanding of the relationships among watershed characteristics and riparian vegetation (chapter 7) provides the basis for developing effective management and restoration scenarios at the watershed/riparian corridor scale. Table 9.1 shows the dominant geomorphic and vegetational characteristics of the watersheds and their relative sensitivities to disturbance.

TABLE 9.1.

Basin sensitivity categories for central Great Basin, the geomorphic characteristics and vegetation types typical of the different categories, and the relative sensitivity of each category to natural and anthropogenic disturbance

	Geology	Geomorphic Characteristics	Vegetation Types	Watershed Sensitivity
Group 1 Flood- dominated channels	Tertiary volcanic rock	• High-relief basins • Narrow bedrock controlled • Minimal sediment storage • Multiple, discontinuous terraces	*Terrace dependent* • *Salix*/forb • Dense rose • *Populus* spp.	Very high
Group 2 Deeply incised channels	Tertiary volcanic rock	• Large, high-relief basins • Incised channels/trenched fans • Multiple, semicontinuous terraces	*Terrace dependent* • *Salix*/grass or forb • *Rosa* • *Artemisia*	Low to moderate
Group 3 Fan- dominated channels	Sedimentary and meta- sedimentary rock	• Lower-relief basins • Large side-valley fans • Metastable channels with low incision values but active downcutting	*Above fans* • Wide riparian zones; Meadow complexes *At and below fans* • Narrow riparian zones; Woody riparian types	Low to moderate
Group 4 Pseudo- stable channels	Intrusive igneous, and sedimentary rock	• Moderate/minor incision • Potential for catastrophic incision via groundwater sapping • Cobbles or smaller bed material • Multiple, discontinuous terraces where incised	*Unincised reaches* • Meadow complexes, *Incised reaches* • Meadow vegetation in trough, *Artemisia* types on upper terraces	Moderate to high

Group 1 Basins: Flood Dominated

Group 1 basins are flood-dominated systems that are underlain by Tertiary volcanic rock and characterized by high-relief basins with moderate to high relative stream power (table 9.1; fig. 9.1). These are narrow, bedrock controlled systems that exhibit minimal sediment storage. During large flood events, there is significant cutting and filling that results in multiple, discontinuous terraces.

The vegetation types within these systems are characterized by disturbance-tolerant species. Low to intermediate terraces are dominated by *Salix* spp./mesic forb, dense rose, or *Populus* spp. vegetation types (table 9.1).

FIGURE 9.1. South Twin River is a watershed that is representative of basins categorized as Group 1, flood-dominated channels. Such basins are highly sensitive to natural and anthropogenic disturbance.

Artemisia spp. and *Pinus monophylla* typically occur on the highest terraces. The geomorphic characteristics associated with the streamside vegetation types include high stream gradients and large channel particle sizes but intermediate stream width/depth ratios and incised channel depths.

The channels within these flood-dominated watersheds are inherently unstable and subject to significant channel change during major runoff events. Consequently, these watersheds are more sensitive to both natural and anthropogenic disturbance than any of the other categorized watershed types. These watersheds require proactive management with minimal disturbance of both the uplands and the riparian corridor. Because of the steep and rocky nature of the watersheds, most are unsuitable for livestock grazing. Roads that were originally located in the valley bottoms have largely been eliminated during successive flood events (e.g., in the South Twin River and Pine Creek basins). New roads should be placed well out of the valley bottoms and follow proper engineering standards (USDA Forest Service 1997), or the watersheds should be designated as roadless. Management and restoration efforts should focus on maintaining natural channel configurations and reestablishing the flood-tolerant, woody riparian vegetation that characterizes these riparian corridors. Natural establishment of woody riparian vegetation appears to occur fairly rapidly after flood events, and near-stream *Betula occidentalis*, *Salix* spp., and *Populus* spp. date primarily to flood events within the past fifty years (Henderson 2001).

GROUP 2 BASINS: DEEPLY INCISED CHANNELS

Group 2 basins are incision dominated and are underlain primarily by Tertiary volcanic rocks. The watersheds have large, high-relief basins and relatively high stream power (table 9.1). The stream channels have eroded completely through the side-valley alluvial fans and have the greatest incision values of all the streams examined (fig. 9.2). These channels have longitudinal profiles that are relatively smooth. They also have a high number of semicontinuous terraces. From a channel-evolution perspective, it appears that these watersheds responded first and most significantly to the current phase of channel instability. Development of pool-riffle sequences and establishment of vegetation on the low terraces indicate that channel morphologies are currently somewhat stable and may be approaching a new equilibrium state.

In these systems, the lower terraces are dominated by *Salix* spp./mesic

FIGURE 9.2. Upper Reese River is a watershed that is representative of basins categorized as Group 2, deeply-incised channels. Such basins are moderately sensitive to natural and anthropogenic disturbance.

meadow or *Salix* spp./mesic forb vegetation types (table 9.1; fig. 9.2). Intermediate terraces typically are characterized by *Rosa*-associated vegetation
types, while the upper terraces have *Artemisia* vegetation types. *B. occidentalis*/mesic meadow and *Prunus/Rosa* types occur locally on intermediate terraces in these systems. The predictive variables for the vegetation
types closely parallel those for the stream categories — relatively low to intermediate gradients, high numbers of terraces, and high, incised channel
depths. Lower gradient reaches with finer-textured stream banks tend to be
dominated by grass and sedge understories in Great Basin vegetation types,
while higher-gradient reaches with coarser bank textures are characterized
by forb understories (Weixelman et al. 1996).

These streams are moderately sensitive to disturbance. The magnitude
of the response to future disturbance will depend on the degree to which
the channels have evolved toward a new stable or equilibrium state. Thus,
channels that still have knickpoints and lack smooth longitudinal profiles
likely will exhibit greater responses to high-flow events than those with
smooth longitudinal profiles. In general, these channels should respond
well to proactive management, especially those that are approaching or
have reached an equilibrium state. Careful placement of roads with respect
to the stream channels will prevent the streams from being diverted onto
the roads. Preventing overgrazing by livestock in both the uplands and the
riparian corridor will help moderate stream flows and provide maximum
stream-bank protection. Although these streams may experience lateral migration during high-flow events, future incision of most of these streams is
likely to be minimal. Water tables of the associated riparian ecosystems
should remain relatively stable and, thus, the site potential of degraded riparian ecosystems can be more accurately determined and effective restoration measures implemented.

Group 3 Basins: Fan Dominated

Group 3 basins are underlain primarily by sedimentary and metasedimentary rocks and are strongly influenced by side-valley alluvial fans (table
9.1; fig. 9.3). Fan-dominated watersheds are generally less rugged, have
lower relief, and exhibit lower relative stream power than watersheds that
support Group 1 and 2 streams. Despite relatively low incision values in
comparison to other basin types, the channels are actively downcutting
and, locally, are among the most dynamic in the study area. In these watersheds, the fans result in unstable longitudinal profiles that are relatively

flat upstream of the fans and steepen significantly where the channels traverse the fans. Relatively coarse alluvium associated with the fans has slowed or prevented incision through many of the fans to date, but the over-steepened segments are inherently unstable. When an episode of incision does occur within a fan segment, the incision migrates up channel into the lower-gradient reach of the valley fill and commonly produces erosion terraces. These systems can be described as metastable. The fans have effectively increased watershed response times to natural and anthropogenic disturbances, but multiple episodes of incision slowly result in progressive entrenchment as the streams work to smooth their longitudinal profiles.

The vegetation types within this group depend on geomorphic position relative to the side-valley alluvial fans (fig. 9.3) (Davis 2000; Korfmacher 2001). Valley segments immediately upstream of alluvial fans are not only relatively flat, but are also fairly wide and often exhibit elevated water tables. In those cases where the upstream segments are associated with elevated water tables, they almost always support meadow complexes. Meadow complexes occur along hydrologic gradients, and the common vegetation types along this gradient are, from wettest to driest, *Carex nebrascensis* meadow, mesic meadow, and dry meadow (Weixelman et al. 1996; Castelli et al. 2000). At those locations where the steams traverse the fans, *Salix* spp. and *Populus* spp. vegetation types dominate. Valley segments that are not influenced by the fans are characterized primarily by *B. occidentalis*/mesic meadow or *Salix* spp. vegetation types.

Group 3 watersheds are the least sensitive to disturbance because of the stabilizing influence of the side-valley alluvial fans. However, management and restoration schemes must recognize that even though Group 3 basins exhibit the slowest rates of incision, and have incised the least, they are fundamentally unstable. The streams will continue to incise wherever dissection through the side-valley alluvial fans can occur. This indicates that management and restoration activities should focus on stabilizing the alluvial fans. Factors contributing to stream incision through alluvial fans include improper road placement and maintenance, and unimproved campsites. Thus, roads and land-use activities in these watersheds should be managed to prevent further destabilization of the alluvial fans. Geomorphic analyses of these fan systems shows that the fan deposits are coarser grained and more poorly sorted than deposits above or below the fans and that the coarse-grained nature of the fans has prevented or reduced the rates of incision through the fans (Miller et al. 2001). This indicates that techniques designed to directly increase the stability of the fans without block-

FIGURE 9.3. Kingston Creek is a watershed that is representative of basins categorized as Group 3, fan-dominated channels. Such basins are least sensitive to natural and anthropogenic disturbance.

ing the movement of resident fish or aquatic organisms, such as armoring, may be highly effective.

GROUP 4 BASINS, PSEUDOSTABLE CHANNELS

Group 4 basins are characterized by channels that typically exhibit minimal downcutting but that have the potential for rapid and catastrophic incision on a localized basis (fig. 9.4). These watersheds are underlain primarily by intrusive igneous and sedimentary rocks and, depending on the basin size, tend to have lower relief and less relative stream power than the other watersheds in the study area. Unincised reaches have channels that are either well defined and set about 1 meter into the valley floor or shallow and flowing across flat valley floors or meadows. Channels within incised reaches exhibit prominent knickpoints whose upstream migration often is facilitated by groundwater sapping. These channels have discontinuous terraces.

Within Group 4 basins, incision can occur very rapidly (days to months) as a result of short-term forcings such as high runoff associated with snowmelt and rainfall (Germanoski et al. 2001). These forcings can be precipitated by natural disturbances like wildfire (Germanoski and Miller 1995), anthropogenic disturbances, or management actions. Group 4 basins are among the most variable in the region as many are either unincised or slightly incised, while others are extensively incised, apparently due to threshold-crossing disturbances or runoff events.

The vegetation within these types of watersheds depends largely on the degree of incision and groundwater availability (table 9.1; fig. 9.4). In areas with constrained water tables, unincised streams tend to flow through *Artemisia tridentata tridentata* or other *A. tridentata* vegetation types and are characterized by a narrow band of herbaceous riparian vegetation. Where elevated water tables occur, the streams flow through riparian meadow complexes characterized by *C. nebrascensis*, mesic, and dry meadow vegetation types. Following incision, the lower terraces are often characterized by meadow vegetation types and the upper terraces by *A. tridentata* types. Streams with well-developed meadow vegetation on the lower terraces, like Illipah Creek, typically have low stream gradients, small channel substrate sizes, fine textured banks, and low width/depth ratios (chapter 7).

Group 4 basins have the potential for rapid and catastrophic incision and, thus, are highly sensitive to both climate and anthropogenic disturbance. Disturbance in the uplands, such as wildfire, also can result in incision when coupled with high-flow events (Germanoski and Miller 1995). This indicates that the uplands should be managed to reduce the risk of

FIGURE 9.4. Indian Creek is a watershed that is representative of basins categorized as Group 4, pseudostable channels. Such basins are highly sensitive to natural and anthropogenic disturbance.

high runoff events due to factors such as widespread overgrazing or large-scale wildfires. Also, vegetation manipulations such as wood cutting or prescribed burning should be restricted to relatively small management units. Roads should be located away from the valley bottoms, and overgrazing by livestock should be prevented. Stream diversion and water spreading should be avoided. Once the basin has been incised, its recovery to the pre-disturbance state is unrealistic, and restoration activities should be based on the current site potential once a new stable state has been established.

Valley-Segment/Stream-Reach Scales

At local scales, like the valley-segment or the stream-reach scales, an understanding of the ecological types, alternative states, and thresholds that currently exist for incised or degraded ecosystems can be used to determine the site or restoration potential and to develop appropriate management approaches (Hobbs and Norton 1996; Whisenant 1999; Hobbs and Harris 2001). The basis for using alternative-state and threshold concepts in the applied sciences is founded on the assumption that multiple alternative states of a given ecological type can exist on the landscape. Ecological types are areas with distinctive climates, landforms, soils, and vegetation. Alternative states exist when different biotic communities occur within a given ecological type. The changes between states are attributed usually to either natural and anthropogenic disturbances or management actions (Westoby et al. 1989; Rodriguez Iglesias and Kothman 1997; Allen-Diaz and Bartolome 1998). New states exhibit different structures and processes that are adjusted to the altered factors or processes (Whisenant 1999). Thresholds exist between the states that can be defined based on the parameters that determine the limits of natural variability within the system (Ritter et al. 1999). Threshold crossings can be recognized when a system does not return via natural processes to the original state following disturbance (Laycock 1991; Ritter et al. 1999). For vegetation communities, threshold crossings can be biotic in nature, resulting from internal interactions such as competition or from external factors such as species invasions. They can also be abiotic, resulting from changes in factors such as weather patterns, geomorphic processes, soils, or hydrologic regimes. If the changes in structural and functional processes are of sufficient magnitude, the result can be a new ecological type with a different set of restoration potentials.

To use alternative-state and threshold concepts in a restoration context, it is necessary to both identify the ecological types and alternative states that exist for the disturbed or degraded ecosystem and define the thresholds between the possible types and states in terms of quantifiable ecosystem characteristics (Whisenant 1999; Chambers and Linnerooth 2001; Wright and Chambers 2002). Defining the thresholds that exist between states is necessary because the recovery potential of a given ecosystem depends on the characteristics of the threshold that exists between the states. If a threshold has not been crossed, then the system should be able to recover to the original state with minimal intervention. Simply eliminating the perturbation, such as overgrazing, or reinstating natural processes, such as flooding, may result in the return of the system to a less-degraded state. If a threshold has been crossed, a new state exists that has a unique set of possible trajectories. It may be possible to return the system to the original state through some type of restoration activity if an abiotic threshold such as a change in soil properties or decrease in water table has not been crossed. If an abiotic threshold has been crossed, in most cases it will be necessary to treat the system as an entirely new state or a new ecological type and to develop the appropriate restoration or management methods for the new state or ecological type.

Riparian Ecological Types for the Central Great Basin

Understanding the ecological types for riparian ecosystems begins by defining the relationships among the geomorphic characteristics, hydrologic regimes, soil attributes, and vegetation types that occur within these systems. A regional classification of riparian ecological types has been developed for the central Great Basin (Weixelman et al. 1996). It identifies recurring ecological types that have distinctive climates, landforms, soils, and vegetation. The classification is based on multiple plots sampled at site or reach scales. Fifteen unique ecological types have been identified, and several variants of these types described. Variables important for differentiating ecological types include elevation in the watershed, fluvial geomorphic surface (floodplain, gravel bar, drainageway, stream terrace, spring, toeslopes), valley slope and lateral slope (perpendicular to valley slope), and soil characteristics (temperature at 50-centimeter depth, texture, soil moisture, depth to field capacity, and depth to soil saturation) (fig. 9.5). The first major division is based on landform, lateral slope, and depth to saturation, illustrating the overriding importance of geomorphic position and water availability in determining ecological site types for these ecosystems (see chapter 7). Vegetation types dominated by grasses and herbaceous species, or *Salix* species with grass and sedge understories, occur on sloping landforms including trough-shaped floodplains and drainage ways and fault-controlled springs with lateral slopes of 5.9 percent ± 5.7 percent (mean ± SD). In contrast, sites with *Salix, Populus, Betula,* or *Artemisia* overstories with either forb or grass and sedge understories are found on steeper landforms such as trough-shaped floodplains, stream terraces, incised land-

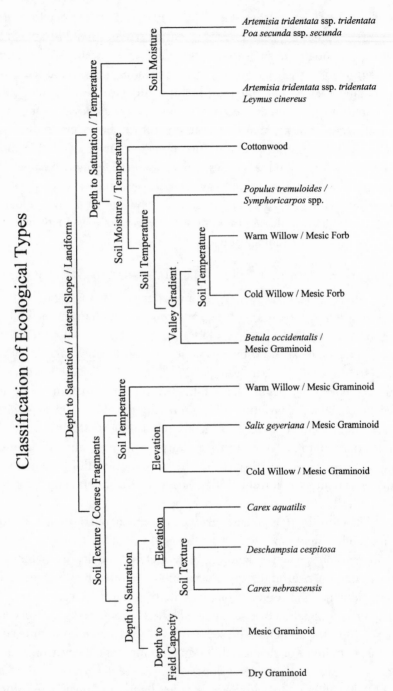

Classification of Ecological Types

FIGURE 9.5. A classification of ecological types for riparian areas of central Nevada that is based on geomorphic and soil characteristics as well as on vegetation composition (from Weixelman et al. 1996).

forms, and toeslopes with average lateral slope of 14 ± 13.5 percent. Soil texture and various indicators of the hydrologic regime (soil moisture, depth to field capacity, and depth to soil saturation) are key variables determining individual ecological types. Elevation within the watershed and soil temperature at 50 centimeters are important because of their effects on species distributions.

ALTERNATIVE STATES AND THRESHOLDS OF MEADOW ECOLOGICAL TYPES

Each of the different ecological types identified for the riparian areas of the central Great Basin can potentially exhibit several different alternative states or plant communities. To date, efforts to develop alternative state and transition models for these ecosystems have focused on riparian meadow ecological types. Riparian meadow complexes occur primarily in incision-dominated watersheds and are subject to high shear stress associated with knickzone migration and processes such as groundwater sapping (Miller et al. 2001; chapters 4 and 5). In the central Great Basin, as elsewhere, the hydrologic regime is the primary determinant of the soil characteristics and the plant community composition and dynamics of riparian meadow complexes (Weixelman et al. 1996; Chambers et al. 1999; Castelli et al. 2000; Martin and Chambers 2001a,b, 2002). Stream incision lowers the base level of groundwater discharge, causes lowered water tables, and results in a change in the composition of the riparian vegetation (chapter 5). Progressive degradation of meadow complexes can result in encroachment and dominance of *A. tridentata tridentata* (basin big sagebrush) into areas formerly characterized by meadow vegetation (Chambers and Linnerooth 2001; Wright and Chambers 2002). Because meadow complexes represent a highly valuable resource and many are at high risk due to ongoing stream incision, they have been the subject of several EM Project studies. At low to intermediate elevations, the common vegetation types along the hydrologic gradient within these meadow complexes are, from wettest to driest, *C. nebrascensis* meadow, mesic graminoid, dry meadow, and *A. tridentata/Leymus cinereus* (Weixelman et al. 1996; Castelli et al. 2000). These meadow ecological types can be characterized by an "average" water table depth and relative variability (fig. 9.6) (J. Chambers, unpublished data). Both seasonal and yearly variability about the average water table depth increases as depth to water table increases, and ecological types with the highest average water tables tend to have the lowest variability (Castelli et al. 2000; chapter 5).

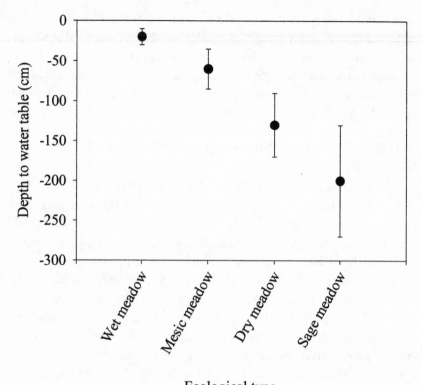

FIGURE 9.6. Water table depths (means ± standard errors) for meadow ecological types typical of intermediate elevations in the central Great Basin.

A hypothetical model of the relationships among the *C. nebrascensis* (wet) meadow, mesic meadow, and dry meadow ecological types, and the potential transitions and alternative states within each of these ecological types, is presented in figure 9.7. The model shows that differences in ecological types are a function of water table regime and that an abiotic threshold determined by water table depth exists between the different types. Alternative states occur within the hydrologic regime (range of water table depths) that characterizes the natural variability of the ecological type.

The alternative vegetation states and transitions within a given ecological type often occur along a disturbance gradient (fig. 9.7). Transitions among phases or plant communities within states are largely reversible unless an ecological threshold has been crossed, but they often require active management. Plant community composition and soil characteristics, such as rates of infiltration and compaction (soil plates), within the different states are useful indicators of ecological status or condition (Weixelman

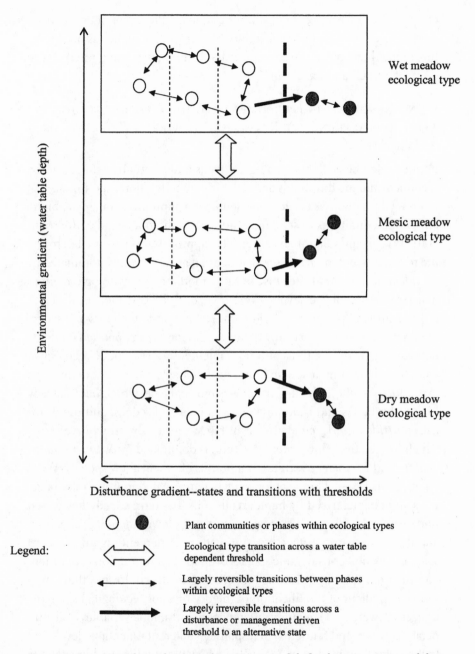

Environmental gradient (water table depth)

Wet meadow ecological type

Mesic meadow ecological type

Dry meadow ecological type

Disturbance gradient--states and transitions with thresholds

Legend:

○ ●   Plant communities or phases within ecological types

⟺   Ecological type transition across a water table dependent threshold

⟷   Largely reversible transitions between phases within ecological types

⟶   Largely irreversible transitions across a disturbance or management driven threshold to an alternative state

FIGURE 9.7. Hypothetical state and transition models for the wet, mesic, and dry meadow ecological types. Differences among ecological types are largely dependent on water table depths; differences among states are largely dependent on disturbance and management history.

et al. 1997). Once an ecological threshold to an alternative state has been crossed, the potential for reversion to the former state depends on the nature of the threshold and resources available for management intervention (Wright and Chambers 2002).

USING RESTORATION EXPERIMENTS TO IDENTIFY ECOLOGICAL TYPES AND ALTERNATIVE STATES AND TO DEFINE THRESHOLDS: A CASE STUDY

For degraded ecosystems, ecological types and alternative states can be identified, and the thresholds defined, through restoration experiments that examine the responses of sites representing an appropriate range of abiotic and biotic conditions to specific treatments (Chambers and Linnerooth 2001; Wright and Chambers 2002). This approach was used to determine the restoration potential of degraded riparian areas currently dominated by A. tridentata tridentata. Because of the importance of hydrologic regimes in determining ecological types (Weixelman et al. 1996; Castelli et al. 2000), it was assumed that the threshold(s) for restoration to the grass-and-sedge-dominated meadows that many of these areas formerly supported could be defined largely on the basis of water table depths. Two model ecological types that share similar soil and landform characteristics were defined for the restoration effort: the dry meadow ecological type, which has relatively high water tables and is dominated by grasses and sedges, and the A. tridentata tridentata/L. cinereus (basin big sagebrush/basin wildrye) type, which has relatively low water tables and is dominated by A. tridentata tridentata and L. cinereus with lesser amounts of dry meadow species (Weixelman et al. 1996). These ecological types are located in trough drainageways, are characterized by haplocryoll (soils that have weakly developed horizons, that have mean annual temperatures lower than 8°C at 50 cm, and that have loamy surface horizons with >2.5 percent organic carbon) soils, and exhibit considerable species overlap. The restoration experiment consisted of applying a burning and seeding treatment to sites that represented a gradient of modification of the dry meadow ecological site type. Objectives were to (1) quantify differences in the abiotic variables (soil temperature, water, and nutrients) in response to different water table levels and to the restoration treatment, (2) evaluate how the abiotic variables affected both seedling establishment and plant regrowth, and (3) use both the abiotic and biotic responses to identify the ecological site types and alternative states and to define the restoration thresholds for these ecosystems.

The study was conducted in the Toiyabe Range in the central Great Basin. The experimental design and methods are detailed in Chambers and Linnerooth (2001) and Wright and Chambers (2002). The experiment included five different *Artemisia*-dominated sites with three water table levels that varied seasonally over the three-year study: (1) wet (–60 to –250 centimeters), intermediate (–100 to –280 centimeters), and dry (–280 to –300 centimeters or more). A wet, intermediate, and dry site were each located in a single drainage; an additional wet site and an additional dry site were located in separate drainages. All of the sites were dominated by *A. tridentata tridentata*. The wet and intermediate water table sites had understories characterized by typical dry meadow species (*Carex douglasii, L. cinereus, L. triticoides, Poa secunda juncifolia,* and *Lupinus argenteus*), and the wet sites had relatively low densities and volumes of *A. tridentata tridentata* (Linnerooth et al. 1998). The restoration treatment involved a paired-plot approach in which one half of each plot served as a control, while the other half was burned to remove the sagebrush and then seeded with native grass and forb species typical of the two model ecological types. The biotic response was evaluated in terms of the seed bank, residual vegetative propagules (root sprouts), seedling establishment, and plant regrowth (Chambers and Linnerooth 2001; Wright and Chambers 2002; Wehking 2002; Blank et al. 2003).

The abiotic and biotic responses of the different sites to the restoration experiment can be used to identify the ecological types and alternative states and to define the thresholds for sagebrush-dominated riparian corridors (fig. 9.8) (Wright and Chambers 2002). Relatively low water tables and abundance of shrubs indicated that the study sites were at the lower end of the water tables necessary to support the dry meadow ecological type. The model dry meadow type is dominated by grasses and sedges and has relatively high depths to saturation (–70 to –100 centimeters in June and July) (Weixelman et al. 1996). Higher water tables facilitate establishment and persistence of species such as *P. secunda juncifolia* that are adapted to mesic conditions and that typify this state. Saturated rooting zones during the spring and early summer prevent establishment and persistence of *A. tridentata tridentata* and other shrubs (Ganskopp 1986). Overgrazing and other perturbations can alter species composition but will not result in shrub encroachment unless the water table drops. Sites with slightly lower water tables (–50 to –250) than the model ecological type, such as the wet and intermediate study sites, are characterized by dry

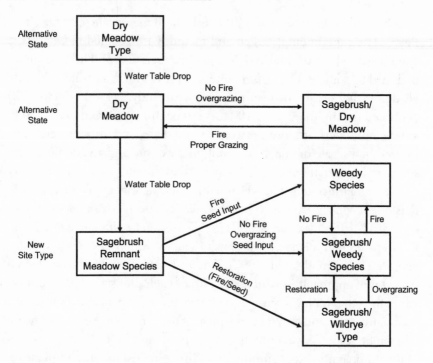

FIGURE 9.8. Diagram of the states and transitions that exist for the dry meadow ecological type of the central Great Basin. Sagebrush = *Artemisia tridentata tridentata*; Wildrye = *Leymus cinereus* (modified from Wright and Chambers 2002).

meadow species but are susceptible to sagebrush encroachment. The processes that allow persistence of the dry meadow species are occasional high water tables coupled with favorable environmental conditions that facilitate episodic establishment of typical dry meadow species (as occurred on our sites in 1998; Chambers and Linnerooth 2001). Water tables remain sufficiently low that *A. tridentata tridentata* can establish and persist (Wright and Chambers 2002). Overgrazing of this state increases sagebrush establishment by reducing competition from the herbaceous species (see Belsky et al. 1999). Once sagebrush has established, fire is necessary to remove the sagebrush and return the state to graminoid and forb dominance. Designation of the wet and intermediate water table sites studied here as an alternative state of the dry meadow ecological type depends on the interpretation of the "range of natural variability" within an ecological type. If the range is considered sufficiently broad to include sites with slightly lower water tables that can exhibit sagebrush encroachment, then the wet and intermediate sites represent alternative states of the dry meadow ecological type.

Sites with the lowest water tables (–250 to more than –300 centimeters) are dominated by sagebrush with a minor component of dry meadow species. These sites have crossed an abiotic threshold and no longer have the potential to support the dry meadow ecological type. The failure of typical dry meadow species to establish (Chambers and Linnerooth 2001) indicates that the dry meadow species that currently occur on the sites may be remnants that existed prior to stream incision and lowered water tables. These sites are currently unstable and can exhibit several different pathways depending on the disturbance regime and initial species composition. The establishment of A. *tridentata* and seeded L. *cinereus* on the dry sites (Wright and Chambers 2002) indicates that these sites have the potential to support at least the more xeric species that characterize the A. *tridentata/ L. cinereus* ecological type (Weixelman et al. 1996). Because the herbaceous understory species that characterize the A. *tridentata/L. cinereus* ecological type are almost nonexistent on the dry sites, their establishment would require active restoration. Fire alone can convert these sites to annual forbs and grasses, especially if the fire-adapted, exotic grass, *Bromus tectorum*, exists in the understory (Wright and Chambers 2002). Depending on the initial species composition, overgrazing and the introduction of invasive species can result in the codominance of A. *tridentata tridentata* and annual species in the absence of fire.

An understanding of the ecological types, alternative states, and thresholds that exist for riparian corridors currently dominated by A. *tridentata tridentata* can be used to design restoration schemes to establish a mosaic of sagebrush and dry meadow ecosystems that more closely resembles predisturbance conditions. Areas that have the potential to support the grass-and-sedge-dominated state can be identified on the basis of water table depths and understory species composition. Restoration that includes prescribed burning and proper management can be used to convert these areas to the grass-and-sedge dominance with minimal intervention and expense. Similarly, sites with the potential to support the A. *tridentata/ L. cinereus* ecological type can be identified. Active restoration techniques including burning, eliminating the invasive annual grasses if present, and seeding with the appropriate species complement can be used to establish this type. The amount of energy required to establish this type is significantly greater than that required for restoring the grass-and-sedge-dominated state of the dry meadow. Economic as well as ecological analyses will be needed to restore the graminoid-dominated state of the dry meadow and establish the A. *tridentata/L. cinereus* ecological type.

Synthesis

The most successful restoration and management approaches are based on knowledge of the underlying processes structuring the systems of interest. The EM Project has used a multiscale approach to obtain the necessary understanding of stream and riparian ecosystem processes to develop effective management and restoration approaches. At the scale of the watershed/riparian corridor, the major focus has been on understanding and categorizing watershed sensitivity to past and present disturbance. Watersheds vary in their sensitivity to disturbance due to differences in controlling factors, such as bedrock geology, basin morphometry, valley width and gradient, and channel/bank substrate. Because the streams in the central Great Basin are currently prone to incision, the categories presented here make use of sensitivity indicator variables that provide insights into the magnitude, timing, and nature of channel incision along the riparian corridors. All of the basins included in the four categories exhibit some degree of channel incision and, thus, the watersheds can be delineated based on variables such as the timing and degree of channel incision, the degree that side-valley fans influence incision, and the presence of geomorphic factors that indicate relative channel stability. The basins fall into four distinct categories: Group 1, flood dominated; Group 2, deeply incised; Group 3, fan dominated; and Group 4, pseudostable. Riparian plant species and vegetation types are closely related to the geomorphic characteristics of the watersheds, and the variables used to categorize the watersheds according to disturbance sensitivity, are useful predictors of riparian vegetation. Combining the categorization of watershed sensitivity with knowledge of the relationships among the geomorphic and vegetation characteristics of the watersheds provides the basis for developing effective management and restoration scenarios. The general approach is applicable elsewhere, but the categories developed use-indicator variables that are specific to the geomorphic responses of upland watersheds in the central Great Basin to disturbance, and it would be necessary to develop indicators specific to the region of interest.

At both the valley-segment and the stream-reach scales, it is often difficult, first, to assess the recovery potential of degraded or disturbed systems (i.e., current versus possible states) and, second, to determine the successional trajectory of the systems once the stressor has been removed or successional processes have been reinitiated via restoration. Alternative-state and threshold concepts can be used for identifying the ecological types and alternative states that exist on the landscape and for evaluating their restoration potential. A critical aspect of the process is defining the thresholds that exist between the states in terms of quantifiable abiotic and biotic ecosystem variables. Restoration experiments that evaluate the responses of sites with the appropriate range of abiotic and biotic conditions can be used both to identify ecological types and alternative states, and to define thresholds. Moreover, this approach can be used to evaluate the potential of the disturbed/degraded system and to design the most effective restoration scenarios.

Acknowledgments

This chapter is the product of the combined efforts of most of the collaborators of the EM Project. We thank Julie Stromberg and Robert Beschta for insightful reviews.

LITERATURE CITED

Allen, E. B., W. W. Covington, and D. A. Falk. 1997. Developing the conceptual basis for restoration ecology. *Restoration Ecology* 5:275–276.

Allen-Diaz, B., and J. W. Bartolome. 1998. Sagebrush-grass vegetation dynamics: Comparing classical and state-transition models. *Ecological Applications* 8:795–804.

Belsky, A. J., A. Matzke, and S. Uselman. 1999. Survey of livestock influences on stream and riparian ecosystems in the western United States. *Journal of Soil and Water Conservation* 54:419–431.

Blank, R. R., J. C. Chambers, and D. C. Zamudio. 2003. Prescribed burning of central Nevada degraded riparian ecosystems: effects on soil and vegetation. *Journal of Range Management* 56:387395.

Castelli, R. M., J. C. Chambers, and R. J. Tausch. 2000. Soil-plant relations along a soil-water gradient in Great Basin riparian meadows. *Wetlands* 20:251–266.

Chambers, J. C., R. R. Blank, D. C. Zamudio, and R. J. Tausch. 1999. Central Nevada riparian areas: Physical and chemical properties of meadow soils. *Journal of Range Management* 52:92–99.

Chambers, J. C., K. Farleigh, R. J. Tausch, J. R. Miller, D. Germanoski, D. Martin, and C. Nowak. 1998. Understanding long- and short-term changes in vegetation and geomorphic processes: The key to riparian restoration. Pp. 101–110 in *Rangeland management and water resources*, edited by D. F. Potts. American Water Resources Association and Society for Range Management. Herndon, Va.

Chambers, J. C., and A. R. Linnerooth. 2001. Restoring riparian meadows currently dominated by *Artemisia* using alternative state concepts—the establishment component. *Applied Vegetation Science* 4:157–166.

Chapin, F. S. III., M. S. Torn, and M. Tateno. 1996. Principles of ecosystem sustainability. *American Naturalist* 148:1016–1037.

Christensen, N. L., A. M. Bartuska, J. H. Brown, S. Carpenter, C. D'Antonio, R. Francis, J. F. Franklin, J. A. MacMahon, R. F. Noss, D. J. Parsons, C. H. Peterson, M. G. Turner, and R. G. Woodmansee. 1996. The report of the Ecological Society of

America Committee on the scientific basis for ecosystem management. *Ecological Applications* 6:665–691.

Clary, W. P., and B. F. Webster. 1989. *Managing grazing of riparian areas in the Inter-mountain Region.* General Technical Report INT-263, USDA Forest Service. Ogden, Utah.

Davis, C. 2000. Influences of alluvial fans on local channel geomorphology, incision and vegetation dynamics in the Toiyabe, Toquima and Monitor Ranges of central Nevada. Master's thesis, University of Nevada, Reno.

Fleischner, T. L. 1994. Ecological costs of livestock grazing in western North America. *Conservation Biology* 8:629–644.

Forman, R. T. T., and R. D. Deblinger. 2000. The ecological road-effect zone of a Massachusetts (USA) suburban highway. *Conservation Biology* 14:36–46.

Ganskopp, D. C. 1986. Tolerances of sagebrush, rabbitbrush, and greasewood to ele-vated water tables. *Journal of Range Management* 39:334–337.

Germanoski, D., and J. R. Miller. 1995. Geomorphic responses to wildfire in an arid watershed, Crow Canyon, Nevada. *Physical Geography* 16:243–256.

Germanoski, D., C. Ryder, and J. R. Miller. 2001. Spatial variation of incision and deposition within a rapidly incising upland watershed, central Nevada. Pp. xi, 41–48 in *Sediment: Monitoring, modeling, and managing.* Proceedings of the 7th Interagency Sedimentation Conference. Vol. 2. Sponsored by the Subcommittee on Sedimentation. Mar. 25–29, 2001. Reno, Nev.

Gerrard, J. 1993. Soil geomorphology, present dilemmas and future challenges. *Geo-morphology* 7:61–84.

Gregory, S. V., F. J. Swanson, W. A. McKee, and K. W. Cummins. 1991. An ecosystem perspective of riparian zones: Focus on links between land and water. *BioScience* 41:540–551.

Harris, R. R. 1988. Associations between stream valley geomorphology and riparian vegetation as a basis for landscape analysis in the eastern Sierra Nevada, California, USA. *Environmental Management* 12:219–228.

Henderson, D. M. 2001. Understanding relationships among climate, stream channel dynamics, and riparian vegetation establishment during the past fifty years in up-land watersheds of central Nevada watersheds. Master's thesis, University of Nevada, Reno.

Hobbs, R. J., and J. A. Harris 2001. Restoration ecology: Repairing the earth's ecosys-tems in the new millennium. *Restoration Ecology* 9:239–246.

Hobbs, R. J., and D. A. Norton. 1996. Commentary: Towards a conceptual framework for restoration ecology. *Restoration Ecology* 4:93–110.

Kauffman, J. B., and W. C. Krueger. 1984. Livestock impacts on riparian ecosystems and streamside management implications . . . a review. *Journal of Range Manage-ment* 37:430–438.

Korfmacher, J. L. 2001. Reach-scale relationships of vegetation and alluvial fans in central Nevada riparian corridors. Master's thesis, University of Nevada, Reno.

Lahde, Daniel B. 2003. Relationships between low-standard roads and stream incision in central Nevada. Master's thesis, University of Nevada, Reno.

Laycock, W. A. 1991. Stable states and thresholds of range condition on North Ameri-can rangelands: A viewpoint. *Journal of Range Management* 44:427–433.

Linnerooth, A. R., J. C. Chambers, and P. S. Mebine. 1998. Assessing the restoration potential of dry meadows using threshold and alternative states concepts. Under-standing long- and short-term changes in vegetation and geomorphic processes: The key to riparian restoration. Pp. 111–118 in *Rangeland management and water resources*, edited by D. F. Potts. American Water Resources Association and Society for Range Management. Herndon, Va.

Martin, D. W., and J. C. Chambers. 2001a. Restoring degraded riparian meadows: Biomass and species responses. *Journal of Range Management* 54:284–291.

———. 2001b. Effects of water table, clipping, and species interactions on *Carex nebrascensis* and *Poa pratensis* in riparian meadows. *Wetlands* 21:422–430.

———. 2002. Restoration of riparian meadows degraded by livestock grazing: Above- and below-ground responses. *Plant Ecology* 163:77–91.

Miller, J. R., D. Germanoski, K. Waltman, R. Tausch, and J. Chambers. 2001. Influence of late Holocene hillslope processes and landforms on modern channel dynamics in upland watersheds of central Nevada. *Geomorphology* 38:373–391.

National Research Council. 1992. *Restoration of aquatic ecosystems: Science, technology, and public policy*. Washington, D.C.: National Academy Press.

———. 2002. *Riparian areas: Functions and strategies for management*. Washington, D.C.: National Academy Press.

Ohmart, R. D. 1996. Historical and present impacts of livestock grazing on fish and wildlife resources in western riparian habitats. Pp. 245–279 in *Rangeland wildlife*, edited by P. R. Krausman. Society for Range Management. Denver, Colo.

Ritter, D. F., R. C. Kochel, and J. R. Miller. 1999. The disruption of Grassy Creek: Implications concerning catastrophic events and thresholds. *Geomorphology* 29:323–338.

Rodriguez Iglesias, R. M., and M. M. Kothman. 1997. Structure and causes of vegetation change in state and transition model applications. *Journal of Range Management* 50:399–408.

Schumm, S. A., and G. R. Brakenridge. 1987. River response. Pp. 221–240 in *North America and adjacent oceans during the last deglaciation: The geology of North America*, edited by W. F. Ruddiman and H. E. Wright Jr. Vol. K-3. Boulder, Colo.: Geological Society of America.

Sidle, R. C., and M. C. Amacher. 1990. Effects of mining, grazing, and roads on sediment and water chemistry in Birch Creek, Nevada. Pp. 463–472 in *Watershed planning and analysis in action*. Proceedings of the Watershed Management Symposium. American Society of Civil Engineers. Jul. 9–11, 1990. Durango, Colo.

Skovlin, J. M. 1984. Impacts of grazing on wetlands and riparian habitat: A review of our knowledge. Pp. 1001–1103 in *Developing strategies for range management*. Society for Range Management. Boulder, Colo.: Westview Press.

Society for Ecological Restoration International. 2002. *The SER Primer on Ecological Restoration*. Science and Policy Working Group. www.seri.org/.

Trimble, S. W., and A. C. Mendel. 1995. The cow as a geomorphic agent: A critical review. *Geomorphology* 13:233–253.

Trombulak, S. C., and C. A. Frissell. 2000. Review of ecological effects of roads on terrestrial and aquatic communities. *Conservation Biology* 14:18–30.

USDA Forest Service. 1997. *The water/road interaction technology series: An introduction*. Technology and Development Program. San Dimas, Calif.

Wehking, P. M. 2002. The role of the seedbank in the restoration of a basin big sagebrush dominated riparian ecosystem to dry meadow. Master's thesis, University of Nevada, Reno.

Weixelman, D., D. Zamudio, K. Zamudio, and K. Heise. 1996. *Ecological type identification and ecological status determination*. R4-ECOL-96-01. USDA Forest Service. Ogden, Utah.

Weixelman, D. A., D. C. Zamudio, K. A. Zamudio, and R. J. Tausch. 1997. Classifying ecological types and evaluating site degradation. *Journal of Range Management* 50:315–321.

Westoby, M., B. Walker, and I. Noy-Meir. 1989. Opportunistic management for rangelands not at equilibrium. *Journal of Range Management* 42:266–274.

Whisenant, S.G. 1999. *Restoring damaged wildlands: A process-oriented, landscape-scale approach.* Cambridge, UK: Cambridge University Press.

Williams, J. E., C. A. Wood, and M. P. Dombeck, eds. 1997. *Watershed restoration: Principles and practices.* Bethseda, Md.: American Fisheries Society.

Wright, M. J., and J. C. Chambers. 2002. Restoring riparian meadows currently dominated by *Artemisia* using threshold and alternative state concepts—aboveground vegetation response. *Applied Vegetation Science* 5:237–246.

ABOUT THE EDITORS AND AUTHORS

Jeanne C. Chambers is a research ecologist with the USDA Forest Service, Rocky Mountain Research Station in Reno, Nevada, and Adjunct Associate Professor at the University of Nevada, Reno. Her primary research focus is on the restoration of disturbed or degraded ecosystems in the western United States. Since 1993, she has served as Team Leader of the Great Basin Ecosystem Management Project for Restoring and Maintaining Sustainable Riparian Ecosystems.

Jerry R. Miller holds the Blanton J. Whitmire Distinguished Professorship of Environmental Sciences at Western Carolina University. He has worked extensively in both the United States and South America on the geomorphic responses of river systems to natural and anthropogenic disturbance and is a coauthor of the third and fourth editions of the textbook *Process Geomorphology* (McGraw-Hill).

Michael C. Amacher is a soil scientist with the USDA Forest Service, Rocky Mountain Research Station in Logan, Utah. He works on soil analysis methods, trace element biogeochemistry in soils, reactivity, and transport of metals in soils, and the status and trend of forest soil quality indicators.

Peter F. Brussard is Professor of Biology at the University of Nevada, Reno, and a codirector of the Nevada Biodiversity Initiative. He has held offices in the American Society of Naturalists, the Society for the Study of Evolution, and the Society for Conservation Biology. His research interests and expertise include conservation biology, ecosystem management, and biogeography of the Great Basin.

Jason B. Dunham is a research fisheries biologist with the USDA Forest Service, Rocky Mountain Research Station in Boise, Idaho. He has worked with a variety of marine and freshwater fishes in tropical and temperate environments. His current research is focused on conservation and landscape ecology of threatened salmonid fishes.

293

Erica Fleishman is Research Associate at the Center for Conservation Biology, Stanford University. Her research focuses on integration of conservation science and land management, particularly in the arid western United States.

Dru Germanoski is Professor and Head of the Department of Geology and Environmental Geosciences at Lafayette College and a registered professional geologist in Pennsylvania. He is a fluvial geomorphologist and his research interests include the relationships between sediment transport and river morphology in gravel-bed rivers, and also landscape and river response to climate change and land-use change.

Paul Grossl is Associate Professor of Biogeochemistry at Utah State University. He is a soil chemist who works on trace element biogeochemistry in soils and methods of remediating contaminated soils to restore soil productivity and decrease health risks.

Kyle House is a research geologist with the Nevada Bureau of Mines and Geology at the University of Nevada, Reno. His research interests include late Cenozoic alluvial stratigraphy, Quaternary paleohydrology, and fluvial geomorphology. His current efforts focus on the use of geologic information for flood hazard assessment on desert piedmonts, the Quaternary evolution of the Truckee River, Nevada, and the late Cenozoic inception and evolution of the lower Colorado River.

David G. Jewett is a hydrologist in the Ground Water and Ecosystems Restoration Division of the U.S. Environmental Protection Agency, Office of Research and Development, National Risk Management Research Laboratory in Ada, Oklahoma. He has more than twenty years of experience in hydrology and hydrogeology, having held positions in academia and industry before joining the U.S. EPA.

John L. Korfmacher is a physical scientist with the USDA Forest Service, Rocky Mountain Research Station in Fort Collins, Colorado. He works on projects that study long-term changes in the hydrology, meteorology, and air and water quality in alpine and subalpine ecosystems. Prior to moving to Fort Collins, he was employed by the Rocky Mountain Research Station's Reno Laboratory, where he worked on studies of the stream systems and riparian areas of the central Great Basin.

Janice Kotuby-Amacher is the Director of Utah State University's Analytical Laboratory. Her research interests include soil analysis methods, nutrient management, soil fertility, and plant and animal nutrition.

Mark L. Lord is a hydrogeologist at Western Carolina University. His areas of interests are hydrogeology, geomorphology, and science education. He has conducted hydrologic and paleohydrologic studies throughout the northern Great Plains, the southern Appalachians, and the central Great Basin.

James A. MacMahon is Trustee Professor in the Department of Biology and the Ecology Center at Utah State University, Logan, Utah. He is an associate editor of *Restoration Ecology*, past president of the Ecological Society of America, a Fellow of the American Association for the Advancement of Science, and a member of the National Research Council's Board on Environmental Studies and Toxicology. He spends his research time equally divided among community studies of plants and animals. Most of his current work involves observing natural systems, and taking lessons from these to be used in applied areas such as restoration ecology.

Scott A. Mensing is Associate Professor in the Department of Geography at the University of Nevada, Reno. He studies Quaternary vegetation and climate change in western North America using pollen and charcoal analysis.

Dennis D. Murphy is Research Professor at the University of Nevada, Reno, and a Pew Scholar in Conservation and the Environment. His research focuses on the biology of butterflies and conservation planning under the federal Endangered Species Act. He currently serves as science advisor to the CALFED Bay-Delta Program, an environmental restoration program for the Sacramento and San Joaquin Rivers.

Cheryl L. Nowak is a paleoecologist with the Reno laboratory of the USDA Forest Service, Rocky Mountain Research Station. She studies the contents of fossil woodrat nests collected from the Great Basin to better understand local and regional plant history and climate change events.

Robin J. Tausch is a range scientist and Project Leader with the USDA Forest Service, Rocky Mountain Research Station in Reno, Nevada. His research focuses on the ecology and paleoecology of Great Basin watersheds with an emphasis on the long-term dynamics of pinyon-juniper woodlands, and their associated sagebrush ecosystems, in response to past and possible future climate change.

David A. Weixelman is a botanist with the USDA Forest Service, Tahoe National Forest in Nevada City, California. He has extensive experience in sampling, analysis, and classification of upland vegetation in Alaska, and riparian vegetation in the Great Basin and Sierra Nevada.